# Sustainability and the Philosophy of Science

This book demonstrates how the philosophy of science can enhance our understanding of sustainability and the practices we use to enact it. Examining assumptions about concepts, theories, evidence, and the moral ideals of sustainability can better orient us as we pursue this urgent and important goal.

The book applies perspectives and tools from the philosophy of science – construed broadly to include portions of science and technology studies, history of science, and philosophy more generally – to sustainability discourse. It argues that widely held assumptions regarding the meaning of concepts, methods of theorizing and inferential practice, evidential structure, and ethics limit our understanding and practice of sustainability. It offers philosophical alternatives that capture more fully the confusing, wicked nature of sustainability challenges. The alternatives draw attention to existing but often undervalued frameworks in sustainability discourse.

This book is aimed towards academics, researchers, and post-graduates working in sustainability, as well as philosophers of science and environmental philosophers interested in the philosophical issues raised by the pursuit of sustainability.

**Jeffry L. Ramsey** is an Associate Professor in the Philosophy Department at Smith College, Northampton, MA, USA. He is also a member of Smith's Programs in Environmental Science & Policy and History of Science & Technology. He teaches courses in the philosophy of science, history of science, environmental philosophy, and the history of early modern European philosophy.

# Routledge Focus on Environment and Sustainability

For more information about this series, please visit: www.routledge.com/
Routledge-Focus-on-Environment-and-Sustainability/book-series/RFES

# Sustainability and the Philosophy of Science

Jeffry L. Ramsey

Routledge
Taylor & Francis Group
LONDON AND NEW YORK

earthscan
from Routledge

First published 2024
by Routledge
4 Park Square, Milton Park, Abingdon, Oxon OX14 4RN

and by Routledge
605 Third Avenue, New York, NY 10158

*Routledge is an imprint of the Taylor & Francis Group, an informa business*

© 2024 Jeffry L. Ramsey

*British Library Cataloguing-in-Publication Data*
A catalogue record for this book is available from the British Library

ISBN: 978-1-032-21503-7 (hbk)
ISBN: 978-1-032-21504-4 (pbk)
ISBN: 978-1-003-26869-7 (ebk)

DOI: 10.4324/9781003268697

Typeset in Times New Roman
by codeMantra

**For Nancy**

# Contents

# Acknowledgments

I owe a debt of gratitude to many people. Numerous students in my Sustainability seminar over the years have provided a sounding board for ideas that eventually made it into this book. Students have also pushed me to think about the topic in new ways; Vinny Farley and Annika Lackner in the Fall 2021 iteration of the class are good examples. Their respective attention to language and power helped me articulate several points in the book. Angie Bottomley served as a research assistant in the same semester, plunging headlong into the literature on coproduction. She went above and beyond the call of duty. Only a small portion of the literature she retrieved and summarized for me is discussed in Chapter 3: Theorizing about Sustainability. The rest awaits a fuller treatment later. My colleagues in the Philosophy Department and in the Program in Environmental Science and Policy at Smith College have served as valuable interlocutors. I would like to mention Alex Barron, Efadul Huq, Susan Levin, and Leslie King in particular. Audiences at Association of Environmental Studies and Sciences conferences in 2016 and 2017 provided friendly and helpful feedback. Responsibility for any errors, of course, rests with me.

Portions of chapters appeared, in different forms, in other venues.

I am grateful to Routledge for permission to incorporate portions of my article commentary "Something Wicked This Way Comes" (Ethics, Policy & Environment, September 2017) in Chapters 3 and 5.

I am grateful to Springer for permission to incorporate portions of my book review "Defining Sustainability" (*Journal of Agricultural & Environmental Ethics*, December 2014) in Chapters 1–3, and for permission to incorporate portions of my article "On Not Defining Sustainability" (*Journal of Agricultural & Environmental Ethics*, December 2015) in Chapters 2 and 3.

# 1   Introduction

Humans have always had an impact on the planet. Over the last 200 years however, widespread industrialization, increasing levels of consumption, population increases and other factors have increased the intensity and scope of the impact. Recent reports continually underscore and magnify the severity of the effects. Climate change, extinction rates above the normal background rates, overexploitation of fisheries and land for food production, ever-greater consumption – all are real and present dangers (cf. United Nations, 2019b, p. 4). Moreover, these are not 'merely' environmental issues affecting non-human parts of the planet. For instance, the loss of biodiversity is "also a development, economic, security, social, moral and ethical issue" (IPBES, 2019). If left unchecked, this loss will erode "the very foundations of our economics, livelihoods, food security, health and quality of life worldwide" (United Nations, 2019a).

Clearly, action is needed to address the impacts. 'Sustainability' (and its close cousin 'sustainable development') is commonly used to conceptualize and practice such action. Sustainability is supposed to allow us to live "without unfairly compromising the options for future generations to make choices about their lives" (S. Levin, 2012, p. 431). However, current formulations are not leading to significant change. As Van der Leeuw et al. (2012) put it, "silence roared" in 2010 when one researcher asked a room of prominent sustainability scientists what sustainability problems had been solved in the last decade. "Sustainable development is eluding the entire planet" (Sachs, 2012, p. 2208).

Nagatsu et al. (2020) argue that a critical viewpoint from the philosophy of science can "contribute to the development and soundness of sustainability science" (p. 1807) *via* analysis of epistemological issues, conceptual questions, and the role of values in the field. In a similar vein, Elliott (2018) argues that the philosophy of science can help environmental policy makers by clarifying the roles that values play in policy-relevant science, illuminating the necessary role of dissent for the creation of objective science and policy, and facilitating alternative problem framings through the use of non-technical, non-formal ways of responding to uncertainty. I fully agree with both claims. However, I think they do not go far enough. Focusing on sustainability *science*

DOI: 10.4324/9781003268697-1

limits the discussion. Sustainability is inherently normative, invoking ideals of nature and society. Analyzing the science necessarily involves addressing larger questions about the relation of science to society. Hence, my starting point includes the broader societal discourse about sustainability. Focusing on the methodological issues mentioned by Elliott likewise limits the discussion by taking the science at face value. I explore a complementary approach that aims to benefit environmental policy making generally and sustainability in particular by developing conceptions of meaning, theoretical structure and inference, evidential structure, and normativity that are better suited to the complex issues involved.

The philosophy of science – construed broadly to include portions of science and technology studies, history of science, and philosophy more generally – engages in several projects. It critiques the ontological, epistemological, and ethical assumptions of scientific proposals, clarifies their concepts, helps formulate new concepts and theories, and fosters dialogs among different sciences and between science and society (LaPlane et al., 2019). For instance, "[t]he imperative to link knowledge to action is widely invoked as a defining characteristic of sustainability research" (West et al., 2019, p. 534; cf. Arnott et al., 2020; Miller, 2012). The philosophy of science can critique assumptions invoked in construals of that link, asking whether they advance our pursuit of sustainability. Part of that critique involves an interrogation of methods of theorizing, modeling, and evidential reasoning, allowing one to question what counts as legitimate science.[1] The critique also involves an investigation of the assumptions about humanity's relation to the natural and social worlds, joining ongoing discussions that question the instrumentalist and technocratic assumptions of mainstream sustainability proposals (Miller, 2015; Norton, 2005, 2015; O'Neill et al., 2008; Thompson, 2010).

Concept clarification and help formulating new concepts are needed because sustainability discourse draws on "abstract concepts from a wide range of disciplines," developing and deploying "them in new and different contexts in which they may well carry different meanings and implications" (Nagatsu et al., 2020, p. 1811). Are the deployments valid? Moreover, many sustainability concepts are thickly evaluative, involving inseparable normative and descriptive elements. What do such concepts look like? When can they be deployed justifiably?

As some of the previous remarks hint at, the philosophy of science can help integrate more fully and substantially the social, cultural, and human dimensions of sustainability. Miller (2015), Maggs and Robinson (2016), Sze (2018) and many others have noted that these dimensions have been systematically marginalized in favor of a science-based conception, as if the problems were independent of human action. Since sustainability is at heart a question about how we and others want to live now and a concern about what future lives will be, it is important to critically examine the values and ethics involved in trying to live more sustainably.

To pursue these philosophical projects in matters of sustainability, philosophical tools that are appropriate to the subject matter are needed. Tools and analytical methods developed in the philosophy of science during the greater portion of the 20th century are, for the most part, not suitable. They were constructed to make philosophy more scientific or to allow instances of science to be reconstructed as instances of pre-existing philosophical theories. Versions of the tools made their way into scientific, other humanistic and even lay discussions. With few exceptions, the narrower and broader discussions unsuccessfully illuminated the epistemological credibility of the sciences, often because the tools required toy versions of finished science. Interpreting the verification of scientific models using philosophical conceptions of confirmation (and falsification) is a well-known example (Oreskes et al., 1994). Similar lessons hold for causal reasoning (Cartwright, 1983, 2007), evidential relations (Lloyd, 2012; Longino, 1990, 2013), theoretical structure and inference (Wilson, 2006, 2017), and reductionistic research strategies (Wimsatt, 2007). In consequence, philosophers of science have developed perspectives that illuminate better the philosophical issues arising in actual, live scientific practice. I use these freely, finding them particularly helpful for addressing the confusing, multi-dimensional, multi-scalar, and normatively thick questions of sustainability.

Using these perspectives, I do not enter conflicts about the 'right' way to conceive of sustainability, arguing *ad passim* that such a conception is probably not possible. Following Longino (2013) and others, I examine proposals in order to understand their epistemological structure, their ontological assumptions, their methodological commitments, and their moral assumptions. This shifts the discussion away from questions surrounding which elements are more foundational and focuses attention instead on how sustainability issues are framed. In turn, this allows me to illuminate more fully the epistemological and normative credibility of specific sustainability claims.

To apply the newer philosophical perspectives, I need a view of sustainability, including its scientific elements, that acknowledges its thickly descriptive and evaluative nature. The precise content of that view is developed as the book progresses, but it can be stated briefly as

To say 'X is sustainable' is to invoke a normatively and descriptively 'thick' model of a wicked problem.

I believe this view provides a more robust means of understanding what our commitments regarding sustainability are and should be. For instance, because model-based science forces one to acknowledge both the partial nature of conceptual representations and the embedded nature of data, the claim places limitations on conceptions of sustainability. Similar strictures apply regarding the meanings of sustainability, the inferences we can make with it, and the ethical commitments involved in pursuing sustainability.

What do I mean by 'model,' 'normativity,' and 'wicked problem,' the central elements of the claim? To say something is a model is to say that any sustainability claim is a *selective* claim. One cannot say 'x is sustainable' without selecting what is being sustained, for whom it is being sustained, for how long it is being sustained, and why it is being sustained (Kates et al., 2005; O'Neill et al., 2008). There are many valid and robust possibilities for each of these four dimensions, so any claim about a 'sustainable X' is one of many possible interpretations of what it required to sustain that entity or process. I explore at length the model-based nature of sustainability claims, examining how this affects our understanding of what sustainability is. Model-based science involves simplifying abstractions and idealizations, trade-offs regarding what is and is not important in describing and explaining the phenomenon, different representational aims, multiple modeling approaches, multiple models with competing assumptions, and disagreement about the significance of a model for understanding the actual phenomena (Potochnik 2012; cf. Cartwright, 1983; Giere, 2010; Longino, 2013; Suarez, 2009; Wimsatt, 1987). Different approaches to these issues affect what we consider to be the boundaries of the problems, their scale, the entities that are 'inside' the problem and those that are 'outside' it, and the relations among the entities (Kay, 2008).

Recognizing the model-based nature of sustainability claims affects what sustainability means, the theories we can construct, the evidence we can gather, and the normative ethical stances we can adopt. For example, evidence claims in sustainability are examples of "complex empiricism," in which the relevance and importance of any given piece of evidence can be recognized *only* by utilizing non-trivial and theoretically informed assumptions that specify particular data as relevant to the claim (Lloyd, 2012; van Fraassen, 2008). Recognizing the model-based nature of sustainability amounts to asserting its descriptively 'thick' nature. One cannot state in advance what sustainability is. Rather, one must begin with the practices that invoke a sense of sustainability and explore what it means in context as one tries to practice sustainability.

Sustainability is shot through with normativity, which I believe adds another dimension of thickness. As the Brundtland definition has it, sustainability involves meeting "the needs of the present without compromising the ability of future generations to meet their own needs" (WCED, 1987, p. 6). Any invocation of "needs" includes (or should include) moral and ethical considerations of what is required in order to live and/or flourish. Other definitions and characterizations are similarly normative, invoking considerations of the quality of life, values, capabilities, virtues, justice, rights, and so on. However, as Jamieson (1998), Thompson (2010) and others have noted, these approaches invoke a 'thin,' nonsubstantive conception of sustainability. In these approaches, saying something is 'sustainable' is no more informative than saying it is 'good.' What is good? Why is it good? In addition, the normative issues are often viewed as just another data point to be added to the discussion. They are taken as pre-existing in the community, needing only to be measured but not justified. Or they are treated as detachable from the scientific matters of sustainability. For instance, DeVries (2012) opens the question about quality of life but then drops the questions about responsibility, dreams, and destinies in order to focus on the physical aspects of sustainability.

In contrast, I follow Norton (2005, 2015) and Thompson (2010) in endorsing a 'thick' sense of the normativity involved in sustainability claims. Such a concept is both evaluative and descriptive.[2] Even more, it is 'thick' because it is both action-guiding and world-guiding. The idea here is that one begins with words that seem to get a purchase on the actions and things we encounter in the world, and which become understandable as we describe and enact them (Kirchin, 2013; Williams, 1985). I interpret this in a loose Wittgensteinian sense. Based on a kind of unarticulated sense of things, a sense drawn from what it means to keep something going, we say that this does or does not count as an instance of the concept. And we give reasons why this is so. This leads to an "expressive-collaborative" rather than "theoretical-juridical" (Walker, 2007) conception of sustainability's normativity. We might appeal to an analytical-theoretical conception of sustainability to help us flesh out what sustainability means in each context, but it is the reasons we give that carry the day ethically and scientifically.

Not all sustainability problems are wicked problems. But most are (Norton, 2005, 2015; Thompson, 2010). They are immensely complex, emerging "at the nexus of ecological, social, economic and cultural aspects as well as of normative and political issues of equity and justice" (Caniglia et al., 2020, p. 93). There are many characterizations of wicked problems, but for my purposes here they are:

1 A "confusing mess of inter-related problems" (Termeer et al., 2013, p. 28) that have multiple possible framings. Different features can be validly cited as underlying conditions, causes, effects, side effects, and/or solutions;
2 Multi-dimensional and multi-scalar. Our ability to know these problems is limited, and the consequences of interventions cannot be known in advance;
3 "Highly resistant to solutions" (Termeer et al., 2013, p. 29) and, as such, cannot be managed by experts alone (Ludwig, 2001);
4 Unique (Rittel & Webber, 1973);
5 Changing over time naturally and as a result of our interventions. This applies to both the natural and human elements of the systems (Clark & Dickson, 2003; Persson et al., 2018; Peter & Swilling, 2014).

Regarding this last point, it is important to note that, given the changes occurring in the situations, it is not merely that different 'takes' by different experts will represent the system differently (and thus that there are multiple valid representations of the problem). It is true that there will be different constructions of the problem by different people. But the last point is stronger: it says that the problem itself changes over time. Any hope of a universal, timeless representation of the problem is unfounded.

Why approach sustainability in this way? First and foremost, it provides a way out of the some of the unhelpful pictures of sustainability. Through a series of ontological, epistemological, methodological, and ethical assumptions, many mainstream sustainability proposals "discipline" (Miller, 2015) the challenges by simplifying them, eliding the complexities at the heart of any sustainability claim. The unhelpful, simplifying assumptions include:

1   A universalistic, theory- or knowledge-first view of the science, in which unconditioned knowledge about what would count as sustainability is produced first to hopefully inform or improve actions (Akamani et al., 2015; Biesbroek et al., 2015; K. Levin et al., 2012; Miller, 2013, 2015; Wiek et al., 2012), or 1a. An "integration imperative" (Klenk & Meehan, 2015) in which diverse, transdisciplinary insights are transformed into a homogeneous whole;
2   Simple empiricism about data and evidence (Maggs & Robinson, 2016);
3   Reductionism (Sterman, 2012);
4   Functionalism about systems in which problems originating in the natural sphere generate 'appropriate' responses in the policy realm (Wellstead et al., 2016) and thus a 'loading-dock' view of science and policy (Dilling & Lemos, 2011; Dow & Carbone, 2007; Jasanoff, 2012); and
5   A "theoretical-juridical" conception of ethics (Walker, 2007) and a similar conception of values that acknowledges normative issues in sustainability but reduces values and ethical principles to uncritically adopted exogenous stances.

I think of this as a 'folk conception' of sustainability. It appeals to common scientific and lay understandings of how science works and how it relates to society. Those understandings are not explicitly false (as is often asserted in philosophical discussions of folk psychology), but they are idealized and abstracted to the point of unhelpfulness. Every mainstream proposal does not involve a commitment to all the assumptions, and there are many alternative approaches that deny one, some, or all of them, but they are commonly employed.

Many have already noted the inaptness and unhelpfulness of the assumptions. I use the philosophy of science to deepen and extend this ongoing discussion. This allows me to advocate for re-shaped relationships among the natural science, social science, and humanistic elements of sustainability problems. This re-shaping does a better job of accounting for the normatively and descriptively thick model-based set of practices and insights that comprise sustainability.

In addition to diagnosing ways in which many mainstream approaches over-simplify and thus mischaracterize the nature of sustainability, approaching sustainability *via* the central claim has several advantages. At the empirical level, we can legitimate and validate several robust, relatively theory-independent characteristics of sustainability problems. I say they are 'relatively theory-independent' because they do not appear as a feature only after a particular theory or framework has been adopted. Whatever one's take, they are common features of sustainability challenges. First, making something more sustainable – or trying to make it so – often results in what Sterman (2012) calls "policy resistance." As we intervene with policies, the system responds by moving to a new configuration where the policy does not work or does not work as well. Sterman (2012) provides several examples: road building programs designed to reduce congestion do exactly the opposite (because we change our behavior in light of the programs); forest fire suppression causes greater tree density and fuel accumulation, leading to larger, hotter, and more dangerous fires; pesticides and herbicides stimulate the evolution of resistant pests as well as kill off natural predators; etc. Second, trade-offs (Hirsch et al., 2010; McShane et al., 2011; Thompson, 2018) feature prominently in sustainability discussions. In a given situation, it is nearly impossible to maximize all our sustainability goals. The systems are too complexly interdependent. Policies designed to stabilize the integrity of a food production system can easily lead to reduced ecological integrity and/or to decreased livelihoods for some participants. Third, sustainability proposals are inherently partial. Given that we try to balance multiple objectives in complex systems that change (on their own and in response to interventions), sustainability is a moving target. Fourth, focusing on sustainability as a model-based, thick concept makes sense of the path dependence of sustainability solutions. Path dependence is a kind of inertial resistance to change even before a change is proposed or enacted. This is often called 'lock-in.' The inertia is driven by favorable initial (social, economic, environmental) conditions and the increasing returns to scale as the system grows (Seto et al., 2016). In the environmental arena, analogs abound to the difficulties of replacing the QWERTY keyboard. Finally, it acknowledges the scalar complications of sustainability. Practices and approaches that might be sustainable at the local level often involve unsustainable elements at larger levels, and *vice versa* (Sassen et al., 2013; Thompson, 2010).

At the methodological level, keeping the normatively and descriptively thick model-based nature of sustainability fully in view keeps us from endorsing the unhelpful simplifications. Many proposals that at first glance seem to endorse complexity do not, in the end, do so fully. For instance, accounts based in socio-ecological systems (SES) recognize the inherent connections among the ecological, economic, and social dimensions of sustainability (S. Levin, 2012; Matson et al., 2016). However, they often employ a concept of a system that is too deterministic and that thus encourages a search for a theoretical view 'from above.' And if they recognize the unknowability, indeterminacy, and complexity as they describe the issues, their solutions invoke an attitude of 'if we just get enough people and enough different sciences, we will understand the problem so that we can hand it off to the policy makers.' As I argue in Chapter 3, others fall into a 'transdisciplinary trap' or an 'interdisciplinary imbroglio,' thinking that an approach that welcomes more voices will find an integrated, unified approach. Even approaches to sustainability that invoke an 'action first,' 'transformative approaches,' or 'solutions first' methodology (cf. Loorbach et al., 2017; Miller, 2015; Van der Leeuw et al., 2012 for characterizations) often implicitly or explicitly endorse many of the ideals of the folk theory of sustainability. They continue to appeal to simple empiricism about evidence, a traditional model of theory structure and inference (even if carried out in an inter- or transdisciplinary mode), and a pluralistic voicing of values. Lastly, many calls to include values involve only calls to recognize them. This is a kind of simple empiricism about values in which values are taken as a given rather than needing justification.

At the philosophical level, characterizing sustainability in this way offers several advantages. First, it provides an opportunity to ask questions whether other approaches are appropriate to sustainability problems and thus whether they can hope to serve as a basis for the action that is needed. Wittgenstein asks us to examine the background, i.e., the actions that contextualize and root action. Are the 'natural' moves the appropriate moves? Few approach matters of sustainability asking this question (Little et al., 2016). Parker (2014) comes close but then uncritically invokes systems theory and critical realism as new philosophical bases that can address the issues. Moreover, Parker leaves the discussion at the level of questions that a philosophy of sustainability should ask and does not directly address issues of theoretical and evidential structure or the nature of ethical deliberation in matters of sustainability. Parker's methodology is representative: most plunge directly into the discussion, assuming they have the needed tools at hand. However, since we are not making much progress (as the reports mentioned earlier make clear), we need a sense of whether the tools we use are adequate to the task. Second, approaching sustainability as a normatively thick model-based view of a wicked situation provides an example of and further justification for a robust pluralist stance in science and the philosophy of science (Kellert et al., 2006; Longino, 2013). If all proposals are partial due to the nature of the problem situations and the theories and evidence that can be applied to them, every proposal should be seen as a partial articulation of the descriptive and prescriptive meanings of sustainability. Last, with respect to the specifically philosophical discussions about sustainability, the approach I advocate allows one to strengthen the conception of sustainability so that we do not have to jettison the concept altogether as necessarily bankrupt (cf. Benson & Craig, 2017) or, alternatively, adopt a radically new conception of science that is problematic and/or inappropriate to the problem and their situations. Regarding the latter, attempts to legitimate a "deeper intuitive awareness" while including "non-human knowledge and rights" (Benessia et al., 2012) bypass issues of policy resistance, trade-offs, and scalar complications by suggesting that answers are close at hand. And they suggest that immediate and intuitive responses need not be critically examined. Likewise, attempts to view sustainability as first and foremost a normative ethical principle rather than a scientific concept (Maggs & Robinson, 2016) throw out the mixed nature of the concept and thus make it difficult to integrate normative and scientific concerns.

My approach is perhaps closest in spirit and content to the work of Bryan Norton (2005, 2011, 2015) and Paul Thompson (2010; Thompson & Norris, 2021; Thompson & Whyte, 2011), both of whom have done much to highlight the philosophical issues involved in matters of sustainability. I follow them in characterizing sustainability as a wicked problem. As well, I follow their analysis of sustainability claims as model-based and value-saturated. However, while both include discussion of policy resistance, trade-offs, and path-dependence in their analysis of what sustainability can be, neither foregrounds them. Their theoretical characterizations tend to treat these as resolvable once a suitable philosophical and scientific characterization is adopted. Both also downplay the inherent partiality of sustainability proposals that arise due to the constantly changing relations among the natural and social elements of the systems involved in a sustainability claim. They opt instead for characterizations that cover the changing relations. As I argue primarily in Chapters 3 and 5, both remain too beholden to what Wilson (2006, 2017) refers to as the 'Theory T' analysis of science and to what Walker (2007) calls the theoretical-juridical approach to values and ethics.

Both approach the scientific side of sustainability in 'Theory T' terms, that is by thinking of theories as a formal structure of already given, statable-in-advance stipulations about the meaning of and relations among a set of concepts. They challenge prevalent sustainability characterizations by switching out the conceptual, theoretical, and evidential bases that structure the discourse. For instance, Norton (2005) constructs "a multidisciplinary, integrative language capable of supporting multidisciplinary public discourse and deliberations" (p. 145). This leads him to adopt adaptive management, the safe minimum standard (SMS) of conservation (a rule that states one should save the resource, provided the costs of doing so are bearable), hierarchy theory and other elements (Norton, 2005). Thompson follows a similar line of argument, borrowing and then extending the notion of functional integrity from ecology so that it applies to "a complex web of social, political, and psychological institutions that reproduce and constrain patterns of human behavior" (Thompson, 2010, p. 229).

While valuable as a way of shifting the *content* of the discourse, this leaves the *kinds* of concepts, theories, evidence, and ethical commitments involved in sustainability claims largely unchanged and unchallenged. It leaves unasked the question whether the structures – and thus the kinds of explanation and evidence that will be used in them – are appropriate and whether there might be better options. It leaves us with a 'thin' or, to be more charitable, 'not suitably thick' conception of sustainability. As an alternative, I follow Wilson

(2006, 2017), John Norton (2003, 2021), Love (2012), and others, characterizing theories as a patchwork of projected and extended meanings for scientific terms. In this patchwork, inferential relations are developed *via* material inferences, i.e., inferential relations based on the specific subject matter. On this view, theoretical claims are much more contingent and robustly, inescapably plural. What will count as sustainability will change from place to place and from time to time.

The case is similar regarding values and ethics. Both Norton and Thompson approach the values and ethics of sustainability primarily in theoretical-juridical terms. On this view, morality is represented as a "surprisingly compact kind of theory or some kind of internal guidance system of an agent . . . It makes morality look as it if consists in, or could be represented by, a compact cluster of beliefs" (Walker, 2007, p. 8). Norton advocates identification and use of community-procedural, weak-sustainability, risk-avoidance, and community-identity values (Norton, 2005). These values are added on to the new, more ecologically sensitive conception of resource use mentioned above. In a similar vein, Thompson invokes the agrarian values of localism, self-realization, and the type of citizenship associated with owning land and grafts those onto the notion of functional integrity. Both seem to suggest that these values are 'rock bottom' values, not subject to negotiation. For both, the values are a cluster of beliefs that will – in conjunction with the altered scientific base – guide us toward sustainability. And so we get a relatively thin conception, where one assembles various elements into an interdisciplinary whole that allows us to proceed without problem.

As others have noted, there are problems with approaching the ethical commitments of sustainability in this way. Zia (2018) notes that Norton's hopefulness about communities expressing the values and thereby coming together in order to back a particular action is, in the cases he has studied, not supported. In my reading of Zia's argument, Norton's conception of how the values are mobilized is too thin, not able to account for the wide range of what is "known, felt, and acted out in moral relations" (Walker, 2007, p. 8). McKenna (2011) makes a similar point re Thompson's proposal, noting that Thompson ignores the gendered moral implications of the agrarian values. In so doing, Thompson's values do not account for what is known, felt, and acted out in the agrarian setting. They are too thin. As I argue in Chapter 5, an expressive-collaborative model of ethical decision-making, in which one does not presuppose specific values but allows values of different sorts to be articulated and justified, can do a better job of justifying the values that are invoked. For instance, if we endorse the agrarian values, we also must articulate and justify the implications they carry for relations among genders.

The question of justification – of the theories, explanations, and evidence and of the ethics and values – most differentiates my proposal from Thompson's and Norton's. I ask not only what conception of science and values might provide a more robust understanding of sustainability. I also inquire into what makes the conception better suited to the task and whether the conception it-self is adequately grounded. Why is this justificatory question important? As I indicated in my remarks above, I believe that if we do not reconceptualize the nature of the theories, explanations, evidence and values and their justifica-tion, we will inevitably slide back into thin or thin-seeming conceptions. This is one of Thompson's criticisms of Norton. He argues that Norton is implicitly committed to an approach that is "grounded in the accounting mentality of economic development," lacking any substantive discussion of what things matter and why (Thompson, 2018, p. 22; cf. Thompson, 2010, pp. 234–255). McKenna makes a similar point regarding Thompson, arguing that an agrar-ian vision that does not consider the larger moral landscape will "repeat the mistakes of the previous version of this vision" (McKenna, 2011, p. 534). In sum, without the substantive discussion, we will not know which things mat-ter in particular places and spaces. And we will not know which things *should* matter to the people and the environments there.

I do believe that the more we are aware of and embrace the normatively thick, model-based nature of sustainability claims, the better we will recog-nize that sustainability is "about more than changing lightbulbs" (Weinstein et al., 2013, p. 4; cf. Biermann et al., 2012). That is, we will recognize better the depth and breadth of the problems and what must be changed to address them. I also hope we will be better positioned to make those deep changes. However, I will not attempt to substantiate this latter claim here. My purpose is to clarify the structure and assumptions of mainstream sustainability dis-course. And I hasten to add that seeing sustainability in this way does not pro-vide an easy way to solve or even manage matters of sustainability. To see any proposal as a solution is, as I argue throughout the book, part of the problem itself. But I do hope that we will see matters a little more clearly, get out of the grip of unhelpful pictures, and possibly position ourselves to move forward.

This book has two audiences. First, philosophers of science – and those in the allied fields of the history of science, and science and technology studies, and general philosophy including especially ethics – should see that sustain-ability discussions prominently feature the conceptual, methodological, and epistemological questions with which they are perennially occupied. Philo-sophical questions about theory structure, the role of models, the nature of evidence, explanation, the value of empiricism and positivism, reductionism, the role of values in science – all are prominent. In addition, they should see that more recent philosophical takes on these topics allow a richer characteri-zation of sustainability.

Sustainability researchers and practitioners should see how sustainability necessarily involves commingled epistemological, ontological, methodologi-cal, and ethical claims. Others recognize this descriptively and normatively thick nature of sustainability claims, so this is by no means a novel stance. But opening up the black boxes of meaning, theorizing, evidence, and ethics al-lows one to justify more fully the need to address sustainability claims in their contexts. Having a better sense of what is and what is possible – by having a better sense of what the science is, how the value claims can be justified, and how sustainability relates to society – will position us better to address what sustainability is and the challenges we face as we try to be more sustainable.

Given these two audiences, I explain concepts and approaches familiar to philosophers that are unfamiliar to sustainability researchers, and *vice versa*. Those who are familiar with the concepts and approaches being discussed can jump ahead as needed.

## Outline of the Chapters

In each of the subsequent chapters I examine how the mainstream conception is invoked in a given aspect of sustainability discourse; argue that the approach is inadequate because it does not fit the epistemological and ontological characteristics of the problems; and then, in concert with existing criticisms within sustainability, utilize resources from the philosophy of science and its allied disciplines to point to a better way of approaching normatively and descriptively thick sustainability. Throughout, I refer to a wide variety of sustainability initiatives to illustrate the claims.

In Chapter 2 "The Meanings of Sustainability" I argue that mainstream approaches to sustainability commonly invoke a referentialist conception of sustainability's meaning and semantic normativity. Such an approach has serious difficulties. An alternative 'thick' account of meaning allows us to understand better how and what 'sustainability' means. When sustainability is interpreted as a thick model of a wicked problem, the phenomena of policy resistance, path dependence and trade-offs emerge as natural features of the situation. Additionally, the thick account offers a richer account of sustainability's semantic normativity. That is, it allows us to understand better when an action should count as 'sustainable.' There will be and can be no agreed-upon, statable in advance meaning of sustainability. Rather, what sustainability means and how we do it well has to rely on contexts within which questions and issues about sustainability already make sense.

Chapter 3 "Theorizing about Sustainability" examines several approaches to theorizing. Proponents pursue various purposes with widely different kinds of theories. As they do so, they reassert – sometimes explicitly, sometimes implicitly and sometimes inadvertently – the folk conception and some of its assumptions. As a result, they struggle to capture an appropriate sense of theory for sustainability. As an alternative, I propose viewing scientific theories as a façade of projected and extended meanings for scientific terms (Wilson, 2006) linked *via* material inferences (Love, 2012; Norton, 2003). This provides better ways of thinking about the theoretical structures of sustainability and the inferences that can be drawn from them. Some coproduction approaches to sustainability theorizing closely resemble this alternative.

Chapter 4 "Evidence for Sustainability" examines the role of evidence in sustainability claims. Many have called for 'evidence-based' sustainability analyses. However, there are no "truth telling machines" in sustainability analyses (Fisher et al., 2010). An empirical piece of information like a sustainability indicator is "not a fact" (NRC, 2007, p. 21). I argue that "complex empiricism" (Lloyd, 2012; van Fraassen, 2008; Winsberg, 2018) characterizes better the nature of evidence in sustainability discussions. This sees data or evidence as a representation or model, produced at the end of a process involving theories, (other) models, considerations of data quality (e.g., the correction of measurement, modeling, and sampling uncertainties), and decision-making by scientists about any one or more of the preceding elements. That is, it sees evidence as always involving a three-place relation between data, claim and background assumptions.

Chapter 5 "Ethics and Sustainability" continues the thickening project into the ethical and moral realm. Sustainability is a normative project about how we want to live now and in the future. Yet many if not most proposals endorse a simplified and inadequate version of moral deliberation in their attempts to understand the ethical commitments involved in sustainability. I argue that most proposals conceive of the ethics in sustainability along "theoretical-juridical" lines (Walker, 2007). This treats ethical deliberation about sustainability as more algorithmic than it can be, and it subtly reinforces the split between scientific or moral experts and the people living in the situations. As a result, it hamstrings attempts to address the normatively thick content of sustainability. An "expressive-collaborative" (Walker, 2007) conception directs us to discover and respond to the "geographies of responsibility" involved in moral matters generally and in sustainability in particular. Like the descriptively normative conception of meaning (discussed in Chapter 2), this conception of moral deliberation directs us to attend to the context before drawing any theoretical insights. Every claim must be articulated and justified. Moral knowledge is produced and sustained within communities. The expressive-collaborative model allows us to identify "what kinds of things people need to know to live according to moral understandings that prevail in (any of) their (possibly multiple) communities or societies," and it supplies "critical strategies and standards for testing whether understandings about how to live that are most credited in a community or society deserve their authority" (Walker, 2007, p. 66).

When authors appeal to notions of resilience to explicate their views on sustainability, I comment on their use of sustainability. Since some see sustainability and resilience as linked objective whereas others see them as separate (Marchese et al., 2018), I leave aside a full discussion of how the enriched perspectives on meaning, theory, evidence, and ethics apply to discussions of resilience (or at least the more theoretically characterized versions of that concept). As Thompson (2018) notes, we should not aim to endorse the resilience of every system: systems such as structural racism were (and are) distressingly resilient. We need a way to argue that we should not endorse the sustainability or resilience of these and other such systems. Whether a model-based, normative, thick perspective on resilience can help us with this is a project for the future.

## Notes

1 I follow Mitchell (2009), who argues that our conception of what counts as legitimate science must change given how science is practiced and how it should be used in policy contexts. However, Mitchell bases her argument on the 'emergence' of properties and behaviors in multi-level, multi-component complex systems. For sustainability, the need for a different conception of science and its relation to society stems less from matters of emergence and more from multiple, often conflicting relations among the multiple goals of sustainability. The systems involved may or may not display emergence in Mitchell's sense. Moreover, sustainability involves more than an inquiry into properties and behaviors of complex systems in ways that can be used in policy contexts. Sustainability is a policy matter, but it is not just a question of applying science to policy. Sustainability is a simultaneously descriptive and normative set of questions about socio-ecological systems and their relation to technology, culture, and society.

2 Philosophers debate how to interpret the notions of 'thick' and 'thin' (Kirchin, 2013). Tappolet (2005) interprets *all* moral concepts as simultaneously evaluative and descriptive: 'thin' prescriptive concepts are less descriptively specific than 'thick' concepts, but they are still descriptive. Whether or not that is ultimately true of all moral concepts, I use it here since sustainability claims involve a mix of descriptive and prescriptive elements.

## References

Akamani, K., Holzmueller, E. J., & Groninger, J. W. (2015). Managing wicked environmental problems as complex social-ecological systems: The promise of adaptive governance. *Springer Geography*, 741–762. https://doi.org/10.1007/978-3-319-18787-7_33

Arnott, J. C., Mach, K. J., & Wong-Parodi, G. (2020). Editorial overview: The science of actionable knowledge. *Current Opinion in Environmental Sustainability*, *42*, A1–A5. https://doi.org/10.1016/j.cosust.2020.03.007

Benessia, A., Funtowicz, S., Bradshaw, G., Ferri, F., Ráez-Luna, E. F., & Medina, C. P. (2012). Hybridizing Sustainability: Towards a new praxis for the present human predicament. *Sustainability Science*, *7*(S1), 75–89. https://doi.org/10.1007/s11625-011-0150-4

Benson, M. H., & Craig, R. K. (2017). *The end of sustainability: Resilience and the future of environmental governance in the anthropocene*. University Press of Kansas.

Biermann, F., Abbott, K., Andresen, S., Bäckstrand, K., Bernstein, S., Betsill, M. M., … & Gupta, A. (2012). Navigating the Anthropocene: Improving earth system governance. *Science*, *335*(6074), 1306–1307. https://doi: 10.1126/science.1217255

Biesbroek, R., Dupuis, J., Jordan, A., Wellstead, A., Howlett, M., Cairney, P., Rayner, J., & Davidson, D. (2015). Opening up the black box of adaptation decision-making. *Nature Climate Change*, *5*(6), 493–494. https://doi.org/10.1038/nclimate2615

Caniglia, G., Luederitz, C., von Wirth, T., Fazey, I., Martín-López, B., Hondrila, K., König, A., von Wehrden, H., Schäpke, N. A., Laubichler, M. D., & Lang, D. J. (2020).

A pluralistic and integrated approach to action-oriented knowledge for Sustainability. *Nature Sustainability, 4*(2), 93–100. https://doi.org/10.1038/s41893-020-00616-z

Cartwright, N. (1983). *How the laws of physics lie.* Oxford University Press.

Cartwright, N. (2007). *Hunting causes and using them: Approaches in philosophy and economics.* Cambridge University Press.

Clark, W. C., & Dickson, N. M. (2003). Sustainability science: The emerging research program. *Proceedings of the National Academy of Sciences, 100*(14), 8059–8061. https://doi.org/10.1073/pnas.1231333100

DeVries, B. (2012). *Sustainability Science.* Cambridge University Press.

Dilling, L., & Lemos, M. C. (2011). Creating usable science: Opportunities and constraints for climate knowledge use and their implications for science policy. *Global Environmental Change, 21*(2), 680–689. https://doi.org/10.1016/j.gloenvcha.2010.11.006

Dow, K., & Carbone, G. (2007). Climate science and decision making. *Geography Compass, 1*(3), 302–324. https://doi.org/10.1111/j.1749-8198.2007.00036.x

Elliott, K. C. (2018). Roles for socially engaged philosophy of science in environmental policy. In D. Boonin (Ed.), *The Palgrave handbook of philosophy and public policy* (767–778). Palgrave Macmillan Cham. https://doi.org/10.1007/978-3-319-93907-0

Fisher, E., Pascual, P., & Wagner, W. (2010). Understanding environmental models in their legal and regulatory context. *Journal of Environmental Law, 22*(2), 251–283. https://doi.org/10.1093/jel/eqq012

Giere, R. N. (2010). *Scientific perspectivism.* University of Chicago Press.

Hirsch, P. D., Adams, W. M., Brosius, J. P., Zia, A., Bariola, N., & Dammert, J. L. (2010). Acknowledging conservation trade-offs and embracing complexity. *Conservation Biology, 25*(2), 259–264. https://doi.org/10.1111/j.1523-1739.2010.01608.x

Intergovernmental Science-Policy Platform on Biodiversity and Ecosystem Services (IPBES) (2019). *Report of the plenary of the intergovernmental science-policy platform on biodiversity and ecosystem services on the work of its seventh session.* https://www.ipbes.net/system/tdf/ipbes-7-10_en_adv_1.pdf.

Jamieson, D. (1998). Sustainability and beyond. *Ecological Economics, 24*, 183–192. https://doi.org/10.1016/S0921-8009(97)00142-0

Jasanoff, S. (2012). *Science and public reason.* Routledge.

Kates, R., Parris, T. & Leiserowitz, A. (2005). What is Sustainable Development? Environment: *Science and Policy for Sustainable Development, 47*, 8-21. https://doi.org/10.1080/00139157.2005.10524444

Kay, J. (2008). Framing the situation: Developing a systems description. In D. Waltner-Toews, J. Kay, J. J. Kay & N. M. E. Lister (Eds.), *The ecosystem approach: Complexity, uncertainty, and managing for sustainability* (pp. 15–36). Columbia University Press.

Kellert, S., Longino, H., & Waters, C. K. (Eds.) (2006). *Scientific pluralism.* University of Minnesota Press.

Kirchin, S. (Ed.) (2013). *Thick concepts.* Oxford University Press.

Klenk, N., & Meehan, K. (2015). Climate change and transdisciplinary science: Problematizing the integration imperative. *Environmental Science & Policy, 54*, 160–167. https://doi.org/10.1016/j.envsci.2015.05.017

Laplane, L., Mantovani, P., Adolphs, R., Chang, H., Mantovani, A., McFall-Ngai, M., … & Pradeu, T. (2019). Opinion: Why science needs philosophy. *Proceedings of the National Academy of Sciences, 116*(10), 3948–3952. https://doi/10.1073/pnas.1900357116

Levin, K., Cashore, B., Bernstein, S., & Auld, G. (2012). Overcoming the tragedy of super wicked problems: Constraining our future selves to ameliorate global climate change. *Policy Sciences, 45*(2), 123–152. https://doi.org/10.1007/s11077-012-9151-0

Levin, S. (2012). Epilogues: The challenge of sustainability: Lessons from an evolutionary perspective. In M. Weinstein & R. Turner (Eds.), *Sustainability science: The emerging paradigm and the urban environment* (pp. 431–437). Springer.

Little, J. C., Hester, E. T., & Carey, C. C. (2016). Assessing and enhancing environmental sustainability: A conceptual review. *Environmental Science & Technology*, *50*(13), 6830–6845. https://doi.org/10.1021/acs.est.6b00298

Lloyd, E. A. (2012). The role of 'complex' empiricism in the debates about satellite data and climate models. *Studies in History and Philosophy of Science Part A*, *43*(2), 390–401. https://doi.org/10.1016/j.shpsa.2012.02.001

Longino, H. (1990). *Science as social knowledge*. Princeton University Press.

Longino, H. (2013). *Studying human behavior; How scientists investigate aggression and sexuality*. University of Chicago Press.

Loorbach, D., Frantzeskaki, N., & Avelino, F. (2017). Sustainability transitions research: Transforming science and practice for societal change. *Annual Review of Environment and Resources*, *42*(1), 599–626. https://doi.org/10.1146/annurev-environ-102014-021340

Love, A. C. (2012). Theory is as theory does: Scientific practice and theory structure in biology. *Biological Theory*, *7*(4), 325–337. https://doi.org/10.1007/s13752-012-0046-2

Ludwig, D. (2001). The era of management is over. *Ecosystems*, *4*(8), 758–764. https://doi.org/10.1007/s10021-001-0044-x

Maggs, D., & Robinson, J. (2016). Recalibrating the anthropocene. *Environmental Philosophy*, *13*(2), 175–194. https://doi.org/10.5840/envirophil201611740

Marchese, D., Reynolds, E., Bates, M. E., Morgan, H., Clark, S. S., & Linkov, I. (2018). Resilience and sustainability: Similarities and differences in environmental management applications. *Science of the Total Environment*, *613–614*, 1275–1283. https://doi.org/10.1016/j.scitotenv.2017.09.086

Matson, P., Clark, W., & Andersson, K. (2016). *Pursuing sustainability: A guide to the science and practice*. Princeton University Press.

McKenna, E. (2011). Feminism and farming: A response to Paul Thompson's the Agrarian Vision. *Journal of Agricultural and Environmental Ethics*, *25*(4), 529–534. https://doi.org/10.1007/s10806-011-9328-0

McShane, T. O., Hirsch, P. D., Trung, T. C., Songorwa, A. N., Kinzig, A., Monteferri, B., Mutekanga, D., Thang, H. V., Dammert, J. L., Pulgar-Vidal, M., Welch-Devine, M., Peter Brosius, J., Coppolillo, P., & O'Connor, S. (2011). Hard choices: Making trade-offs between biodiversity conservation and human well-being. *Biological Conservation*, *144*(3), 966–972. https://doi.org/10.1016/j.biocon.2010.04.038

Miller, T. R. (2012). Constructing sustainability science: Emerging perspectives and research trajectories. *Sustainability Science*, *8*(2), 279–293. https://doi.org/10.1007/s11625-012-0180-6

Miller, T. R. (2015). *Reconstructing sustainability science*. Routledge.

Mitchell, S. (2009). *Unsimple truths: Science, complexity, and policy*. University of Chicago Press.

Nagatsu, M., Davis, T., DesRoches, C. T., Koskinen, I., MacLeod, M., Stojanovic, M., & Thorén, H. (2020). Philosophy of science for sustainability science. *Sustainability Science*, *15*, 1807–1817. https://doi.org/10.1007/s11625-020-00832-8

National Research Council (NRC). (2007). *Models in environmental regulatory decision making*. The National Academies Press. https://doi.org/10.17226/11972.

Norton, B. (2005). *Sustainability: A philosophy of adaptive ecosystem management.* University of Chicago Press.

Norton, B. (2011). The ways of wickedness: Analyzing messiness with messy tools. *Journal of Agricultural and Environmental Ethics, 25*(4), 447–465. https://doi. org/10.1007/s10806-011-9333-3

Norton, B. (2015). *Sustainable values, sustainable change: A guide to environmental decision making.* University of Chicago Press.

Norton, J. (2003). A material theory of induction. *Philosophy of Science, 70*, 647–670. https://doi.org/10.1086/378858

Norton, J. (2021). *A material theory of induction.* University of Calgary Press.

O'Neill, J., Holland, A., & Light, A. (2008). *Environmental values.* Routledge.

Oreskes, N., Shrader-Frechette, K., & Belitz, K. (1994). Verification, validation, and confirmation of numerical models in the earth sciences. *Science, 263*(5147), –646. https://doi.org/10.1126/science.263.5147.641

Parker, J. (2014). *Critiquing sustainability, changing philosophy.* Routledge.

Persson, J., Johansson, E. L., & Olsson, L. (2018). Harnessing local knowledge for scientific knowledge production: Challenges and pitfalls within evidence-based sustainability studies. *Ecology and Society, 23*(4), 38. https://doi.org/10.5751/ es-10608-230438

Peter, C., & Swilling, M. (2014). Linking complexity and sustainability theories: Implications for modeling sustainability transitions. *Sustainability, 6*(3), 1594–1622. https://doi.org/10.3390/su6031594

Potochnik, A. (2012). Feminist implications of model-based science. *Studies in History and Philosophy of Science Part A, 43*(2), 383–389. https://doi.org/10.1016/j. shpsa.2011.12.033

Rittel, H. W., & Webber, M. M. (1973). Dilemmas in a general theory of planning. *Policy Sciences, 4*(2), 155–169. https://doi.org/10.1007/bf01405730

Sachs, J. D. (2012). From millennium development goals to sustainable development goals. *The Lancet, 379*(9832), 2206–2211. https://doi.org/10.1016/ S0140-6736(12)60685-0

Sassen, M., Sheil, D., Giller, K. E., & ter Braak, C. J. F. (2013). Complex contexts and dynamic drivers: Understanding four decades of forest loss and recovery in an east African protected area. *Biological Conservation, 159*, 257–268. https://doi. org/10.1016/j.biocon.2012.12.003

Seto, K. C., Davis, S. J., Mitchell, R. B., Stokes, E. C., Unruh, G., & Ürge-Vorsatz, D. (2016). Carbon lock-in: Types, causes, and policy implications. *Annual Review of Environment and Resources, 41*(1), 425–452. https://doi.org/10.1146/ annurev-environ-110615-085934

Sterman, J. D. (2012). Sustaining sustainability: Creating a systems science in a fragmented academy and polarized world. In M. P. Weinstein & R. E. Turner, R. (Eds.), *Sustainability science: The emerging paradigm and the urban environment* (pp. 21–58). Springer.

Suarez, M. (Ed.) (2009). *Fictions in science: philosophical essays on modeling and idealization.* Routledge.

Sze, J. (Ed.) (2018). *Sustainability: Approaches to environmental justice and social power.* New York University Press.

Tappolet, C. (2005). Through thick and thin: Good and its determinates. *Dialectica, 58*(2), 207–221. https://doi.org/10.1111/j.1746-8361.2004.tb00297.x

Termeer, C., Dewulf, A., & Breeman, G. (2013). Governance of wicked climate adaption problems. In J. Kneiling & W. Leal Filho (Eds.), *Climate change governance* (pp. 27–39). Springer-Verlag.

Thompson, P. (2010). *The agrarian vision*. University Press of Kentucky.

Thompson, P. (2018). Norton and sustainability as such. In S. Sarkar & B. Minteer (Eds.), *A sustainable philosophy: The work of Bryan Norton* (pp. 7–26). Springer.

Thompson, P., & Norris, P. (2021). *Sustainability: What everyone needs to know*. Oxford University Press.

Thompson, P. B., & Whyte, K. P. (2011). What happens to environmental philosophy in a wicked world? *Journal of Agricultural and Environmental Ethics, 25*(4), 485–498. https://doi.org/10.1007/s10806-011-9344-0

United Nations (2019a). *"Nature's dangerous decline 'unprecedented': Species extinction rates 'accelerating'*. https://www.un.org/sustainabledevelopment/blog/2019/05/nature-decline-unprecedented-report/

United Nations (2019b). *Global environment outlook 6: Summary for policymakers*. Cambridge University Press.

van der Leeuw, S., Wiek, A., Harlow, J., & Buizer, J. (2012). How much time do we have? Urgency and rhetoric in sustainability science. *Sustainability Science, 7*(S1), 115–120. https://doi.org/10.1007/s11625-011-0153-1

Van Fraassen, B. (2008). *Scientific representation: Paradoxes of perspective*. Oxford University Press.

Walker, M. U. (2007). *Moral understandings: A feminist study in ethics*, 2nd ed. Oxford University Press.

Weinstein, M. P., Turner, R. E., & Ibáñez, C. (2013). The global sustainability transition: It is more than changing light bulbs. *Sustainability: Science, Practice and Policy, 9*(1), 4–15. https://doi.org/10.1080/15487733.2013.11908103

Wellstead, A., Howlett, M., & Rayner, J. (2016). Structural-functionalism redux: Adaptation to climate change and the challenge of a science-driven policy agenda. *Critical Policy Studies, 11*(4), 391–410. https://doi.org/10.1080/19460171.2016.1166972

West, S., van Kerkhoff, L., & Wagenaar, H. (2019). Beyond "linking knowledge and action": Towards a practice-based approach to transdisciplinary sustainability interventions. *Policy Studies, 40*(5), 534–555. https://doi.org/10.1080/01442872.2019.1618810

Williams, B. (1985). *Ethics and the limits of philosophy*. Harvard University Press.

Wilson, M. (2006). *Wandering significance*. Oxford University Press.

Wilson, M. (2017). *Physics avoidance*. Oxford University Press.

Wimsatt, W. (1987). False models as means to truer theories. In N. Nitecki & A. Hoffman (Eds.), *Neutral models in biology* (pp. 23–55). Oxford University Press.

Wimsatt, W. C. (2007). *Re-engineering philosophy for limited beings: Piecewise approximations to reality*. Harvard University Press.

Winsberg, E. (2018). *Philosophy and climate science*. Cambridge University Press.

WCED (The World Commission on Environment and Development) (1987). *Our common future*. Oxford University Press.

Zia, A. (2018). Adaptive management in social ecological systems – Taming the wicked? In S. Sarkar & B. Minteer (Eds.), *A sustainable philosophy – The work of Bryan Norton* (pp. 167–187). Springer Cham. https://doi.org/10.1007/978-3-319-92597-4

# 2 The Meanings of Sustainability

Sustainability is vague. A speaker once bet an audience that there were about 2000 definitions of sustainability in a room of, one supposes, about 2000 people (White, 2013). More commonly, one hears that there are about 300 definitions (Santillo, 2007). The multiplicity is widely viewed as a problem. The presence of so many definitions leaves us "as confused at the end . . . as at the beginning" (Dobson, 1996, p. 33). "Random conceptualizations that do not respect the fundamental sustainability principles undermine the concept's objective to steer action, is self-defeating and arguably inhibits its practical realization" (Waas et al., 2011, p. 1638). And since sustainability involves an appeal to the sciences, precise conceptualizations are suppposed to develop and clarify the content and structure of theories; allow hypotheses to be validated; and allow meaningful evidence to be collected.

In reacting to the problem of vagueness, sustainability proponents commonly employ elements of what is known as the referentialist theory of meaning. Following Wilson (2006) and Cartwright (2020), I argue here that there is good reason to think this theory misconstrues how concepts, especially polysemic, model-based concepts like sustainability deployed to address wicked problems, come to have meaning. First, I argue that the reference of sustainability is settled by use rather than *a priori* stipulation. Second, I argue that its semantic normativity, i.e., whether a use counts as correct or incorrect, is decided by appeal to the irredeemably social nature of rule following. Together, these arguments support a claim that the meaning of sustainability will be vague and contested. As we address a wicked problem with partial, model-based representations of knowledge and action, we select some aspects of the situation as more important and we act out sustainability in multiple ways that connect to different modes of being and acting in the world. This is a first step toward 'thickening' up the concept of sustainability.

## The Reference of Sustainability

In attempting to characterize 'sustainability,' authors commonly invoke a number of elements of the referentialist theory of meaning. Three core elements

DOI: 10.4324/9781003268697-2

are relevant to the discussion of sustainability. First, the theory assumes that singular, or definite, terms acquire meaning by being referred directly to objects. For example, one points to a box and utters the word 'box.' The acts of ostension – the pointings – *explain* why objects are called what they are called by giving a sort of natural history of how the association came about (Harris, 1986). Second, referentialism holds that general or abstract concepts "can be analysed into their characteristic marks (Merkmale) until one arrives at unanalysable simple concepts. . . Analysis terminates with simple unanalyzable concepts such as 'red,' 'dark,' 'sweet'. . . These are explained by ostensive definition" (Hacker, 1975, p. 269). This leads to "the belief that the world is sorted into sharply distinguished kinds, that individual things have essences whose necessary and sufficient conditions we can list in neatly formulated definitions" (Sluga, 2011, p. 76). Finally, the referentialist approach often assumes that meaning is created in the mind of one person and transferred to the mind of another *via* acts of ostension. For example, the teacher's idea of 'box' is transferred to the student by pointing at the object.

These elements are commonly found among proponents of sustainability definitions. Many definitions attempt to make the term specific by providing singular terms in the *definiens* that can be pointed to. For instance, the Brundtland definition (WCED, 1987) clarifies "sustainable development" by reference to "needs," "compromise," "future generations," and a host of subsidiary concepts such as economic growth, equity and environment. 'Need' and 'growth' are supposedly objective, epistemically accessible features of the situation. Likewise, the 'three pillars' and 'three concentric circles' approaches offer characterizations of economic viability, environmental protection and social equity that are intended to be obvious and unproblematic. Other referential strategies such as the UN's Sustainable Development Goals seek to display meaning by making the concept measurable. Each goal is associated with objective indicators that allow us to measure whether we are making progress toward the goal.

Definitions also attempt to characterize the general concept. The approaches mentioned in the previous paragraph are attempts to offer a set of necessary and sufficient conditions, however broad, for use of the concept. Many authors reveal their adherence to referentialism when they complain about the lack of a good definition. Given so many definitions, sustainability is "one of the least meaningful and most overused words in the English language" (Owen, 2011, p. 246). It "lacks solid meaning" (Lutz Newton & Freyfogle, 2005, p. 24). It is "regarded either as internally self-contradictory (an oxymoron) or, at best, plagued by ambiguous or distorted definitions" (Santillo, 2007, p. 60). According to many, the proper response to this situation is to "define [sustainability] clearly" (Thiele, 2013, p. 2), to "reclaim the essence of the initial definition. . ." (Santillo, 2007, p. 61). One needs to

> distill the potent essence of the term [because] it is hard to have a serious
> discussion about a subject as important as sustainability when we don't

even agree on what it means. Harder still if we don't even know that we don't agree on its meaning ("Reviews," n.d.).

Agreeing on the meaning allows us to succeed in "demystifying sustainability" ("Reviews," n.d.).

Finally, many offer definitions so as to transfer the meaning of sustainability from their (expert) heads to others interested in and affected by the matter. Authors define and characterize sustainability so we can learn "what more needs to be done" (Morse, 2010, p. 18). They lay out "a general framework to help people interested in mobilizing knowledge to promote sustainability" (Matson et al., 2016, p. vii). Even if the contested, local nature of sustainability is recognized, somewhat *a priori* principles are enunciated that all interested in sustainability should follow (cf. Jacques, 2021). The UN's Division for Sustainable Development (DSDG) in the United Nations Department of Economic and Social Affairs (UNDESA) claims that "(i)n order to make the 2030 Agenda a reality, broad ownership of the SDGs must translate into a strong commitment by all stakeholders to implement the global goals. DSDG aims to help facilitate this engagement" (SDGs).

The surprise is that this commonly held "picture of the essence of human language" (Wittgenstein, 1958, §11)[1] is seriously inadequate as an account of how words acquire meaning. Wittgenstein argues that "an ostensive definition [for singular terms] can be variously interpreted in *every* case" (Wittgenstein, 1958, §28, emphasis in original).[2] To see this, Wittgenstein asks us to conceive of a "complete primitive language" in which a builder calls out "block," "pillar," "slab," or "beam" to an assistant. The builder calls out a word, and "B brings the stone which he has learnt to bring at such-and-such a call" (Wittgenstein, 1958, §2). Wittgenstein argues that such a language is not sufficient to produce understanding, i.e., knowledge of what the term applies or refers to. If the ostensive teaching has the effect of getting the assistant to bring the appropriate piece,

> am I to say that it effects an understanding of the word? Don't you understand the call "Slab!" if you act upon it in such-and-such a way? – Doubtless the ostensive teaching helped to bring this about; but only together with a particular training. With a different training the same ostensive teaching of these words would have effected a quite different understanding.
>
> (Wittgenstein, 1958, §6)

That is, the meaning of the call "Slab!" is not determined by the words. It is determined by the training that links words and actions, and what the link means "is determined by the practice or custom in which the sign is embedded" (Williams, 2010, p. 80). Whether the call is treated as a judgment ("That is a slab"), a request ("Bring me a slab"), a directive ("Do a dance with that thing!") or something else altogether is set by the 'particular training' in which they are experienced. Wittgenstein calls the training a 'language game.'

Learners cannot learn to make the connections on their own. "Precisely because our behavior is not fully determined innately, the regularities that are necessary to human life can only be normative regularities that are socially sustained" (Williams, 2010, p. 106; cf. Stern, 2009, p. 189). Learning the meaning requires already existing patterns of action and interaction with others. Those patterns allow learners to try out interpretations. Because the regularities are not absolutely hard and fast, new interpretations and mistakes on the part of the learner sometimes fail and sometimes succeed.

How does this apply to sustainability discourse? In the language game of sustainability, most if not all of the *definiens* terms can and are variously interpreted. A few examples will have to suffice.[3] Many have noted the difficulty of interpreting the 'needs' referred to in the Brundtland definition. 'Needs' can refer to needs essential to survive or to satisfaction of needs (i.e., desires) in order to participate in (economic) development (cf. Verburg & Wiegel, 1997). Next, in many definitions of sustainability, "biophilia" is invoked as if it had obvious meaning. However, the "innate tendency to focus on life and lifelike processes" (Wilson, 1984, p. 1) can be interpreted in a variety of ways. 'Innate' can mean one broad instinct or a complex set of such; the 'tendency' can be instinctive or learned, exhibited in all conditions or only contingently in optimal conditions; one can 'focus' *via* attraction, aversion, indifference, peacefulness, anxiety, etc.; and what counts as 'life and lifelike processes' is open to interpretation (Levy, 2003). Additionally, whether one is responding to biological or non-biological (e.g., water, soil, and landscape features) of the environment is open to interpretation. Finally, whether our preference for savannas, which is used to explain why we have the kinds of tendencies we do in fact have, is an adaptive trait itself, an adaptive byproduct of particular psychological adaptations, or an adaptation (of whatever sort) to the physical environment or some combination of the physical and social environment is underdetermined (Joye & de Block, 2011).[4] The same follows for operationalizations. Appeal to sustainability indicators is a common way to supply determinacy to the simple terms (Kates et al., 2005). But there are hundreds of indicators and indices as well as hundreds of indicator initiatives (Wu &Wu, 2012).

Given that there are multiple interpretations of the basic terms, can we say that any of them are correct or incorrect? One can quite plausibly and charitably view the attempts to define sustainability as attempts at instruction regarding proper and improper use. But the definition itself is not sufficient for normativity. That must come from the context of use, in the supplying of alternative interpretations and instruction on when mistakes have been made. It is in use that we ascertain whether the word is used in a way other than we wish (Wittgenstein, 1958, §29). For example, one cannot just supply an interpretation of 'needs' or 'biophilia.' One must argue that the interpretation will do the work needed in each context and should count as a proper meaning. However, one rarely if ever gets any discussion about how the simple terms are to be interpreted. For authors to use terms as if they have fixed definitions is more than simply being careless. "[A] move in chess doesn't consist simply in moving a piece in such-and-such a way on the board. . . but in the circumstances that we call 'playing a game of chess,' 'solving a chess problem,' and so on" (Wittgenstein, 1958, §33). It would be much better to acknowledge that there can be alternatives and thereby fill out what the moves mean. Additionally, unlike the game of chess, we don't yet have an agreed-upon normative structure in the sustainability discussion. We are not yet at the point at which alternatives can be closed off. That is, there is no one who is a master of the language. Moreover, if we merely stipulate the meaning we have not shown why that interpretation is important, i.e., why it should carry any normative force. Ostensive definition of the terms in the *definiens* omits the context of why we are interested in sustainability. Through a discussion of interpretations, we will come to understand more fully how those terms function in what we want from sustainability. For authors to offer definitions using terms as if they fixed meaning and thus proper use of the terms is premature.

Being able to point at the component terms of sustainability requires a social context. What of the general term or concept of sustainability itself? Once we have supplied context for the component terms, couldn't we then have a meaningful general definition of sustainability? Couldn't we fix the 'essence' of sustainability in this way?[5] The Wittgensteinian answer is again no. Because of the social nature of meaning, there are no essences to concepts. And because there are no essences, the meaning of the concept and the understanding of the meaning have to be set by social practices.

Wittgenstein argues that the desire for an essence assumes that meaning exists only when a set of necessary and sufficient conditions has been enunciated. But meaning arises in other ways. Here, Wittgenstein's analysis of concept structure as a set of "family resemblances" is useful. When one looks at games (board games, ball games, card games, etc.), "you will not see something that is common to all, but similarities, relationships, and a whole series of them at that. . . And the result of this examination is: we see a complicated network of similarities overlapping and criss-crossing. . ." (Wittgenstein, 1958, §66). The existence of the complicated network does not lead to meaninglessness. We understand what a concept means by being given examples: "[A]lways ask yourself: how did we learn the meaning of this word. . . ? From what sort of examples? in what language-games? Then it will be easier for you to see that the word must have a family of meanings" (Wittgenstein, 1958, §77).

Thinking of the meaning of sustainability in terms of language games is useful. Although Wittgenstein applies the analysis to the notion of language specifically, "many other concepts must also be family resemblance concepts" (Sluga, 2011, p. 79). 'Sustainability' has an open texture much like 'game,' so I believe it makes sense to view sustainability as a family resemblance concept. To say that there is no set of agreed-upon necessary and sufficient conditions for sustainability is, to put it mildly, an understatement. Many (cf. Dobson, 1996; Jacobs, 1999; Mebratu, 1998) have noted that proponents variously interpret what, why, and who is being sustained. This is nicely illustrated in the National Resource Council's (1999) publication *Our Common Journey*, a canonical if nonetheless controversial sustainability document. The NRC's Board on Sustainable Development noted that organizations and individuals differ on:

- what is to be sustained (nature, life support or community);
- what is to be developed (people, the economy or society);
- the relationship between the two ('only,' 'mostly,' 'but,' 'and,' or 'or'); and
- for how long (25 years, 'now and into the future,' or forever).

Even more, there are differing interpretations of the central terms in each of the first two options. For instance, if nature is to be sustained, are we concerned about sustaining the earth, biodiversity or ecosystems? If people are to be developed, should we focus on child survival, life expectancy, education, equity, or equal opportunity? Additional variations beyond these abound. For instance, sustaining nature can mean an acceptance or rejection of environmental limits (Jacobs, 1999). If resources are the focus in discussions of life support, one can ask whether human, critical, or irreversible capital should occupy our attention (Dobson, 1996).

Given this variability, asking for examples in order to see the complicated network is crucial. Understanding the examples *as examples of* sustainability comes from the context, not a definition. For general terms just as for singular terms, we must rely on a background of taken-for-granted interpretations and actions. That reliance is provisional. Whether an already accepted or proposed new action should count as 'sustainable' comes from "a grasp on things which, although quite unarticulated, may allow us to formulate reasons and explanations when challenged" (Taylor, 1999, p. 32). We try to figure out whether the action 'fits.'

In sum, legislating the reference of sustainability will not come from the definitions alone. Rather, we should look at supposed examples of sustainability. Based on a kind of unarticulated sense of things, a sense drawn from what it means to keep something going, we say that this does or doesn't count. And we give reasons why this is so. We might appeal to the components of a definition to help us out, but it is the reasons we give that carry the day and not the definition itself.

## The Semantic Normativity of Sustainability

It is common to think that if we have 'solid meaning' for sustainability, we can use it differentiate a truly sustainable action from one that is "faux" (Farley & Smith, 2020, p. 4) or a "green or sustainable smoke screen" (Waas et al., 2011, p. 1638). That is, we will be able to characterize the concept's semantic – as distinct from its ethical – normativity. The boundaries of true from false sustainability are commonly seen as arising intellectually, by relying on principles of some sort to set the limits of proper use, or behaviorally, by appealing to real or hoped-for regularities in our behaviors. Here again, the Wittgensteinian argument is that trying to find the semantically normative content of sustainability through either approach misses the mark. Normativity does not arise *via* intellectual or social stipulation. The normativity of sustainability is grounded transcendentally in the irredeemably social nature of rule following, which includes learning and recognizing what counts as an instantiation of the rule.

Examples of the intellectualist approach are numerous. Research in sustainability science has focused largely on understanding complex coupled human-natural systems (Miller et al., 2013). Donella Meadows (1998) is an early, prominent proponent of this approach. "Once we see the relationship between structure and behavior, we begin to understand how systems work, what makes them produce poor results, and how to shift them into better behavior patterns" (p. 1). A process or practice is understood as sustainable by referring to the structures and behaviors, which allow us to see how and when our actions are or are not leading to good lives for all people (aka development) in harmony with nature (aka sustainability). More broadly, sustainability science investigates "those aspects of our understanding of human systems, environmental systems and their interactions that are useful for helping people achieve sustainability goals" (Clark, 2010, as cited in Miller, 2012). Knowledge about the coupled systems differentiates true from faux sustainability.

Another example is Farley and Smith's (2020) notion of neo-sustainability. They articulate three "rules" of neo-sustainability: limits, environmental primacy and systems thinking. Because it invokes the concept of 'carrying capacity,' the rule of environmental primacy is of special relevance to my arguments. This rule is characterized as recognizing that the limits to growth are "dictated by the environment, and therefore actions in any system must adhere to the carrying capacity of the earth's natural system" (Farley & Smith, 2020, p. 175). Whether a process or procedure is sustainable is determined by paying attention to limits to growth and carrying capacities dictated by the first and second laws of thermodynamics. Adopting such a definition reveals that our current economic and social systems violate these rules. The rules tell us when actions are truly sustainable and when they are 'faux.'

Regularism about the normativity of sustainability can be found in several places. Proceduralists commonly ground the correctness of sustainability in behavioral regularities. The National Resource Council (1998) characterizes sustainability as a "process of social learning and adaptive response" (p. 48). John Robinson, who is sometimes referred to as "Mr. Sustainability," defines sustainability as "the emergent property of a discussion about desired futures that's informed by some understanding of the ecological, social, and economic consequences of different courses of action" (Robinson, quoted in Miller, 2012, p. 284). Other proceduralists are even more explicit about the social character of the agreement. "[M]y opinion is that what is sustainable is defined by the stakeholders that will be involved in this process" (Rotmans, quoted in Miller, 2012, p. 284). For proceduralists, sustainability just is what we agree it is in a properly participatory and democratic discussion.

Practice theory approaches to sustainability can also lapse into regularism and thus elide the question about the normativity of meaning. Pfister, Schweighofer, and Reichel (2016) are an example. Practice theory is a sociological-philosophical theory that examines how human action relates to the myriad cultural and social structures in which it exists. Agents make meaning as they enact and alter practices, i.e., routinized, structured behaviors such as cooking, driving, or working. Since the authors rely on Wittgenstein's (1958, §43) dictum that "the meaning of a word is its use in language," one might initially think that they are aware of questions about the normativity of meaning. However, while conceding that sustainability is essentially contested, all of the contestation they analyze revolves around already shared assumptions about what sustainability means. Take their discussion of agricultural food production. In their analysis of the origins of alternative understandings of food production to modern industrial agriculture, they characterize ecological agriculture – in all its instantiations – as "promoting agricultural practices that 'care' for nature rather than just exploit it" (Pfister et al., 2016, p. 69). There is tension among various practices of ecological agriculture, but it is all tension around the meaning of care. 'Good' or 'correct' sustainability just is what people who want to practice 'care' do.

Some proposals combine intellectualist and regularist accounts. Norton (2005) is a good example. The account is more procedural in Norton (2015), but mostly the same structure is present. On this account, sustainability involves at least the following elements: adaptive management; not reducing the ratio of opportunities to constraints from one generation to the next; a safe minimum standard approach to managing catastrophic risk; normative insight; hierarchy theory; discourse ethics; recognition of community-procedural, weak sustainability, risk avoidance and community identity values; and demand modeling, a kind of mission-oriented science that is motivated by a social problem and in which success or failure is measured by whether the models improve the decision process. There is a clear mix of theoretical insight (safe minimum standards, hierarchy theory) and appeals to how people behave (what people see as opportunities, their community values). Overall, the tendency probably tilts toward regularism. Commitments to particular cultural and natural legacies "should reflect the love and respect for stuff that is expressive of the communal values and aspirations that form a community in a place" (Norton, 2005, p. 335). Sustainability is just about what people in particular communities value, clarified by use of the intellectual tools.

Wittgenstein offers two arguments for thinking that these accounts mis-construe the semantic normativity of words in general and sustainability in particular. The first is most relevant to the intellectualist approach:

> [W]e teach [the pupil] to write down . . . series of the form "0, n, 2n, 3n, etc." at an order of the form "+n"; so at the order "+1" he writes down the series of natural numbers. – Let us suppose we have done exercises and given him tests up to 1000.
>
> Now we get the pupil to continue a series (say +2) beyond 1000 – and he writes 1000, 1004, 1008, 1012.
>
> We say to him: "Look what you've done! . . . You were meant to add two: look how you began the series!" . . . [S]uppose he pointed to the series and said: "But I went on in the same way." . . . In such a case we might say, perhaps: It comes natural to this person to understand our order with our explanations as we should understand the order: "Add 2 up to 1000, 4 up to 2000, 6 up to 3000 and so on.
>
> (Wittgenstein, 1958, §185)

The basic negative point is that rules are not self-interpreting. That is, given only a rule, it is always possible to follow the rule in 'deviant' ways that, when explained, are not deviant at all. This process is recursive. Any specification of the rule "would itself be another rule open to deviant application" (Rouse, 2007, p. 502). The basic positive point is that "[t]here is a way of grasping a rule which is not an interpretation, but which is exhibited in what we call 'obeying a rule' and 'going against it' in actual cases" (Wittgenstein, 1958, §201). To get people to do it our way – the 'right' way – we train them in a way that allows them to recognize that the rule applies here.

Wittgenstein reverses the direction of normativity. According to the intellectualist, understanding or knowing allows the correct following to occur. For Wittgenstein, the correct following is shown by exhibiting right and wrong applications.

What does this mean for the intellectualist accounts of sustainability? In all cases, there will be many means of 'going on in the same way.' Remember that the *Our Common Future* definition of sustainability leaves open the interpretation of what is to be sustained, what is to be developed, and for how long. As well, if one chooses to sustain 'nature,' it leaves open what aspect of nature is to be sustained. There are many ways of 'going on in the same way.'

As another example, take the notion of 'carrying capacity' invoked by Farley and Smith. They opt for a broadly ecological interpretation by invoking the "carrying capacity of the earth's natural systems" (Farley & Smith, 2020, p. 175) revealed through ecological footprint analysis. In this sense of 'carrying capacity,' different people might follow the rule by using different assumptions about: the natural states and productive capacities of the environment; what counts as the 'environment' for particular populations; how populations access and use resources in the environment; and so on. Other people might follow the rule by saying, "In this situation, we will use carrying capacity as it is used in theoretical and applied ecology." Here, carrying capacity is used at different times to refer to a population in which the birth rate equals the death rate simpliciter, a population in which the rates equal each other due to density-dependent processes, an average population size that is stable over time due to the combined effects of birth and death rates along with immigration and emigration, or the size that can be maintained by reference to the resource that is in the shortest supply (Hixon, 2008). Further, different people might link carrying capacity to different kinds of footprint analysis, as different analyses can use different assumptions about, among other things, the relative environmental impacts of land use by infrastructure vs. agriculture, the relation of the environmental footprint to sustainability, and the effects of technological change (Van den Bergh & Verbruggen, 1999). Last but not least, the pupil might understand the rule as applying one way in biological contexts but another way in cultural contexts, where considerations of the quality of life that is experienced, often through technological and political innovations, are considered important (cf. Aiken, 1980). Farley and Smith (2020, p. 21) are aware of some of these multiple interpretations, but they simply assume that their interpretation is the one everyone will follow to delineate true from false sustainability.

Wittgenstein's second argument about normativity of meaning applies more directly to the regularist account.

> Let me ask this: what has the expression of a rule–say a sign-post–got to do with my actions? What sort of connexion is there here?–Well, perhaps this one: I have been trained to react to this sign in a particular way, and now I do so react to it.
>
> But that is only to give a causal connexion; to tell how it has come about that we now go by the sign-post; not what this going-by-the-sign really consists in. On the contrary; I have further indicated that a person goes by a sign-post only in so far as there exists a regular use of sign-posts, a custom.
>
> (Wittgenstein, 1958, §198)

Regularists provide only a causal, historical connection between behaviors and the judgment that they are right or wrong. That is, they tell us only that people have connected these actions and assessments, not *why* a particular connection should be regarded as stipulating correct usage. Further, they provide no way of elucidating the boundaries of what can count as a 'right' or a 'wrong' action (cf. Medina, 2003).

For sustainability, this point is clear. If, as Rotmans noted earlier, sustainability is just what "is defined by the stakeholders that will be involved in this process," then anything can be sustainable and any action will be just as 'truly' sustainable as any other. Pfister et al.'s (2016) practice theory analysis of sustainability as a 'care' attitude faces a similar problem. As I noted earlier, they focus on the conflict of interpretations within already given normative framings of sustainability.[6] But what if an agreed-upon ecological agricultural practice is not the way to go on? Attention to this is needed because practices of ecological agriculture are not normatively correct in and of themselves. Because they are embedded in larger social, political and natural systems, actions of 'care' can exhibit path dependence and lead to policy resistance, both of which can lead toward unsustainability initially and over longer periods of time. Take the president of Sri Lanka's 2021 almost immediate overnight ban on synthetic fertilizers and pesticide imports (Torella, 2022). The ban seems to align with the practice of ecological agriculture since it was imposed to reduce the amount of spending on synthetic fertilizer, lessen adverse health and environmental impacts, and support the country's traditional agricultural food production systems. However, the ban caused a drop in the production of rice and tea. The loss in production meant the government had to spend more money on rice imports than would have been spent on fertilizer imports. The government also had to spend millions to compensate farmers for the loss of tea production. In short, the system exhibited policy resistance. That resistance probably came at least in part from path dependencies among the agricultural and economic sectors. Sri Lanka had become self-sufficient in rice production due to the use of synthetic fertilizers and pesticides, a significant achievement from a food-security standpoint. With no immediate alternative to that set of practices, yield dropped. The drop was likely exacerbated by the fact that many had left the agricultural sector during the rise in production, and workers were not available to farm the land ecologically.

More generally, practices alone do not stipulate sufficiently the boundaries of sustainability. Interpreting ecological agriculture as a 'care' practice, as Pfister, Schweighofer, and Reichl do, doesn't mark out good vs. bad sustainable agriculture. For many, stewardship is a form of caring that is significantly different than a purely economic approach to resource management. As Mathevet, Bousquet, and Raymond (2018) point out, there are at least four different construals of stewardship in sustainability science and conservation biology. Each offers differing approaches to the role of science, to recognizing and responding to the plurality of values, and to the capacity to modify values, rules and decision procedures. As another example, think of 'greenwashing.' While some greenwashing involves outright lying, much revolves around practices that others view as not doing enough for sustainability or as being still outright unsustainable. All actions save those involving lying are aiming toward some reduction of socio-ecological impact, but claiming the practices aren't sustainable isn't enough. We need more than the practices to decide whether they count as sustainable. In short, practices don't provide a sense of 'how to go on.'

If the intellectualist and regularist accounts fall short of delineating correct from incorrect interpretations, how does one get semantic normativity? Wittgenstein's insight about the normativity of custom or regular use is cryptic, but the overall import can be discerned. As noted earlier, it helps to think in terms of students and learning. Teachers help students recognize when a behavior has been learned correctly (Williams, 2010). My favorite example here is music lessons. The teacher has to somehow show that the interpretation is acceptable. This involves words, actions, corrections, etc.

Thinking about the nature of practices themselves also provides insight. Practices are mutually interactive, involve something at issue and at stake in the outcome, and pertain to stakes that are perspectivally variant or open textured (Rouse, 2007). All of these are normative. Interaction among constitutive performances identifies the bounds of a practice. Wittgenstein focuses on language. Again, I like to think of music performance: whether a playing is an acceptable playing of that particular piece depends on the context of the performance and the history of performing that piece. In the context of sustainability, one has a sort of call-and-response structure in which an action is proposed as sustainable. The community articulates a sense that a performance does or does not count by referring the proposal to the larger discussion about why we wish to become more sustainable. That is, they articulate it *vis-a-vis* the things at issue and at stake. In everyday communication, the meanings of words are constituted by values of wanting our actions and utterances to make sense to others, wanting communication among members to be smooth, and believing it is a virtue to pursue truth. Here, "we can see that our meaning what we do has at least three kinds of normative import" (Horwich, 2019, p. 153). Other meta-values, i.e., values about the point of the activity, are invoked in the language game of sustainability, but the point is the same. Normativity arises due to larger considerations than just behaving as others do. Lastly, normativity arises because what is at stake is open textured. In matters of sustainability as in many other social practices, we want many things and are faced with complex, interactive systems. We develop models to give us perspectival views about the situation. In such circumstances, performances of practices are "ongoing patterns of causal intra-action within partially shared circumstances" (Rouse, 2007, p. 52). There are no 'ultimate' meanings of any given performance of sustainability, so we check in with each other and constitute the meaning as we do so.

This may sound like regularism, but there are important differences. For one, rules emerge out of practices rather than being laid down to regulate practices.

> We do not first grasp meanings [of social practices] in the head and then try to attach them to publicly recognizable marks. . . . From those articulative capacities and our shared vulnerability and neediness, languages and other social practices become pervasive features of our world. Our

participation in those practices enables us to become the agents we are through our mutual accountability to the possibilities those practices make available and to what is thereby at stake for us in how we respond to those possibilities.

(Rouse, 2007, p. 53)

Second, we don't agree by majority vote to treat proposed rules as norms. Wittgenstein makes a transcendental argument in which practices provide the conditions for the contexts of inquiry and justification (Medina, 2003). Like Kant, Wittgenstein inverts the usual way of approaching the issue. Just as Kant argues that we must assume space and time in order for us to have sensations in space and time, Wittgenstein argues that we must assume the practices in order for us to be able to detect and name regularities as 'regular' in them. Practices supply the conditions for the possibility of the rules and thus make semantic normativity possible.

It might also sound very much like pragmatism in its appeal to community experience. But pragmatism is often regularist in its approach to values. It takes the values that communities express as given, with little exploration about what counts as good or reprehensible courses of action (Loman, 2020; Michael, 2016). On the approach I am suggesting, the values have to be adjudicated with respect to their purposes, the way they come to express normative visions of responsible actions and good communities. It means working out in concrete situations how values are expressive of the good life of a community. How this works is the subject of Chapter 5.

## Implications

In conclusion, we cannot define or characterize sustainability *a priori*. Meaningful definitions and characterizations, ones that people can follow and implement, presuppose a social context in which one can give definitions. Offering a definition and hoping it will be filled out with meaning by people acting in particular ways is just thinking that the meaning of the term will jump magically from the head of the theorist to those in the community. For meaning to happen, the social practices of interpreting have to already be in place.

One might object that one should just point and insist on an interpretation. But this just leaves open the question of what the pointing refers to and why it should be interpreted in that way. The meaning of words in general and of sustainability in particular do not arise full-blown in the mind of a definer and get passed to observers solely *via* concatenations of pointings and sayings. Absent a pre-existing social context of use, legislated, imposed meanings will fail to have any definitive connotation and cannot serve as a guide in deciding whether the word or concept is being applied correctly. We are not yet in a settled language game where a particular interpretation is the only game in town. Any specific interpretation is and can only be defended against a background of interpretation that is already shot through with meaning and contains the possibility of objections to particular uses. One might also object that since we rely on social practices, anything goes and we have been led into relativism. I do not think so. As my remarks about normativity indicate, we determine correct from incorrect usages by trying out alternatives. Alternatives are always set against a background, and the language game in the context of that background will allow us to say that a term has been misinterpreted or has not found a place in the game. This may not be as determinate as many would like, but it does not mean ascriptions of proper and improper use are impossible.

My arguments imply that sustainability will always be contested, not just because it is a political concept (Connelly, 2007; Jacobs, 1999), but because whether a word is necessary in a definition

> depends on whether without it the other person takes the definition otherwise than I wish. And that will depend on the circumstances under which it is given, and on the person I give it to. And how he takes the definition is seen in the use that he makes of the word defined.
>
> (Wittgenstein, 1958, §29)

There are and will be many circumstances under which the concept of sustainability is given, many people to whom it is given, and many uses of the concept. This is especially true in the complex, wicked problems of sustainability, where no one possesses a bird's eye view of the situation and, moreover, our interventions keep changing the system. Rather than using definitions and characterizations to get rid of contestation, we should learn to live with it.

And we will need to live with it in a radical sense. The arguments about how descriptive and normative meaning arise that I have offered do not provide a direct answer about what sustainability means. While this can be frustrating to people, myself included, who want to link knowledge to action in matters of sustainability, I believe it is necessary. We should always ask ourselves whether the assumptions we make are appropriate to the situation. Without doing so, we risk offering solutions that will not move us in the right direction.

## Notes

1   I follow the standard practice of referring to passages in the Philosophical Investigations by the numbered remark in which they appear.

2   Wittgenstein explores how ostension is supposed to work in two related but slightly different situations: ostensive definition and ostensive teaching. Since the issues involved in ostensive definition require a settled language community (Williams, 2010) and since we do not yet have that for sustainability, I omit discussion of this context here.

3   See Ramsey (2015) for a brief discussion of various interpretations of interconnect-edness, interdependence, carrying capacity, biophilia, and entropy, each of which is featured in many definitions of sustainability.
4   The quality of the evidence for each of the interpretations varies quite widely, as do the implications of each (Levy, 2003). However, these issues go beyond the point that the term is interpreted in a variety of ways.
5   Searching for essences is often phrased as laying out the necessary and sufficient conditions for the use of a concept. I am unclear whether sustainability theorists are invested in the specific desire to provide a set of necessary and sufficient conditions for sustainability. However, they are heavily invested in the analytical project of differentiating correct from incorrect usage by fixing the meaning of the concept.
6   Admittedly, this level of discussion might be beyond the scope of the Key Ideas series that their book is published in. Nonetheless, the retreat to regularism is worrisome.

## References

Aiken, W. (1980). The "carrying capacity" equivocation. *Social Theory and Practice,* *6*(1), 1–11. https://doi.org/10.5840/soctheorpract19806114

Cartwright, N. (2020). Middle-range theory. *Theoria: An International Journal for Theory, History and Foundations of Science, 35*(3), 269–323. https://doi.org/10.1387/theoria.21479

Connelly, S. (2007). Mapping sustainable development as a contested concept. *Local Environment, 12*(3), 259–278.

Dobson, A. (1996). Environment sustainabilities: An analysis and a typology. *Environmental Politics, 5*(3), 401–428. https://doi.org/10.1080/09644019608414280

Farley, H. M., & Smith, Z. A. (2020). *Sustainability: If it's everything, is it nothing?* Routledge.

Hacker, P. M. (1975). Wittgenstein on ostensive definition. *Inquiry, 18*(3), 267–287. https://doi.org/10.1080/00201747508601765

Harris, J. F. (1986). Language, language games and ostensive definition. *Synthese, 69*(1), 41–49. https://doi.org/10.1007/bf01988286

Hixon, M. (2008). Carrying capacity. In S. E. Jorgensen & B. Fath (Eds.), *Encyclopedia of ecology* (pp. 528–530). Newnes.

Horwich, P. (2019). Wittgenstein (and his followers) on meaning and normativity. *Disputatio. Philosophical Research Bulletin, 8*(9), 147–172. https://doi.org/10.5281/zenodo.2652686

Jacobs, M. (1999). Sustainable development as a contested concept. In A. Dobson (Ed.), *Fairness and futurity* (pp. 21–46). Oxford University Press.

Jacques, P. (2021). *Sustainability: The basics,* 2nd ed. Routledge.

Joye, Y., & De Block, A. (2011). "Nature and I are two": A critical examination of the biophilia hypothesis. *Environmental Values, 20*(2), 189–215. https://doi.org/10.3197/096327111x12997574391724

Kates, R., Parris, T. & Leiserowitz, A. (2005). What is sustainable development? *Environment: Science and Policy for Sustainable Development, 47*, 8–21. https://doi.org/10.1080/00139157.2005.10524444

Levy, S. S. (2003). The biophilia hypothesis and Anthropocentric environmentalism. *Environmental Ethics, 25*(3), 227–246. https://doi.org/10.5840/enviroethics200325316

Loman, O. (2020). A problem for environmental pragmatism: Value pluralism and the sustainability principle. *Contemporary Pragmatism, 17*(4), 286–310. https://doi. org/10.1163/18758185-17040003

Lutz Newton, J., & Freyfogle, E. (2005). Sustainability: A dissent. *Conservation Biology, 19*, 23–32. (2017). *Sustainability*, 157–166. https://doi.org/10.4324/9781315241951-14

Mathevet, R., Bousquet, F., & Raymond, C. M. (2018). The concept of stewardship in sustainability science and conservation biology. *Biological Conservation, 217*, 363–370. https://doi.org/10.1016/j.biocon.2017.10.015

Matson, P., Clark, W. & Andersson, K. (2016). *Pursuing sustainability: A guide to the science and practice*. Princeton University Press.

Meadows, D. (1998). *Indicators and information systems for sustainable development*. The Sustainability Institute. https://donellameadows.org/archives/indicators-and-information-systems-for-sustainable-development/.

Mebratu, D. (1998). Sustainability and sustainable development: Historical and conceptual review. *Environmental Impact Assessment Review, 18*(6), 493–520. https://doi. org/10.1016/S0195-9255(98)00019-5

Medina, J. (2003). Wittgenstein and nonsense: Psychologism, Kantianism, and the habitus. *International Journal of Philosophical Studies, 11*(3), 293–318. https://doi. org/10.1080/0967255032000108020

Michael, M. (2016). Environmental pragmatism, community values, and the problem of reprehensible implications. *Environmental Ethics, 38*(3), 347–366. https://doi. org/10.5840/enviroethics201638329

Miller, T. R. (2012). Constructing sustainability science: Emerging perspectives and research trajectories. *Sustainability Science, 8*(2), 279–293. https://doi.org/10.1007/s11625-012-0180-6

Miller, T. R., Wiek, A., Sarewitz, D., Robinson, J., Olsson, L., Kriebel, D., & Loorbach, D. (2013). The future of sustainability science: A solutions-oriented research agenda. *Sustainability Science, 9*(2), 239–246. https://doi.org/10.1007/s11625-013-0224-6

Morse, S. (2010). *Sustainability: A biological perspective*. Cambridge University Press.

National Research Council. (1999). *Our common journey: A transition toward sustainability*. National Academies Press.

Norton, B. (2005). *Sustainability: A philosophy of adaptive ecosystem management*. University of Chicago Press.

Norton, B. (2015). *Sustainable values, sustainable change*. University of Chicago Press.

Owen, D. (2011). *The conundrum: How scientific innovation, increased efficiency, and good intentions can make our energy and climate problems worse*. Riverhead Books.

Pfister, T., Schweighofer, M., & Reichel, A. (2016). *Sustainability*. Routledge.

Ramsey, J. (2015). On not defining sustainability. *Journal of Agricultural and Environmental Ethics, 28*(6), 1075–1087. https://doi.org/10.1007/s10806-015-9578-3

Reviews (n.d.) of H. Washington (2015). *Demystifying sustainability*. Routledge. http://www.routledge/com/books/details/9781138812697.

Rouse, J. (2007). Practice theory. In S. Turner & M. Risjord (Eds.), *Philosophy of anthropology and sociology* (pp. 499–540). North-Holland.

Santillo, D. (2007). Reclaiming the definition of sustainability. *Environmental Science and Pollution Research - International, 14*(1), 60–66. https://doi.org/10.1065/espr2007.01.375

SDGS. *Do you know all 17 SDGs?* https://sdgs.un.org/goals.

Sluga, H. (2011). *Wittgenstein*. Blackwell.

Stern, D. (2009). Wittgenstein's critique of referential theories of meaning and the paradox of ostension, philosophical investigations §§26–48. In E. Zamuner & D. Levy (Eds.), *Wittgenstein's enduring arguments* (pp. 179–208). Routledge.

Taylor, C. (1999). To follow a rule. In R. Schusterman (Ed.), *Bourdieu: A Critical Reader* (pp. 29–44). Blackwell.

Thiele, L. (2013). *Sustainability*. Polity Press.

Torella, K. (2022). Sri Lanka's organic farming disaster, explained. *Vox*. https://www.vox.com/future-perfect/2022/7/15/23218969/sri-lanka-organic-fertilizer-pesticide-agriculture-farming.

van den Bergh, J. C. J. M., & Verbruggen, H. (1999). Spatial sustainability, trade and indicators: An evaluation of the 'ecological footprint.' *Ecological Economics, 29*(1), 61–72. https://doi.org/10.1016/s0921-8009(99)00032-4

Verburg, R. M., & Wiegel, V. (1997). On the compatibility of sustainability and economic growth. *Environmental Ethics, 19*(3), 247–265. https://doi.org/10.5840/enviroethics199719314

Waas, T., Hugé, J., Verbruggen, A., & Wright, T. (2011). Sustainable development: A bird's eye view. *Sustainability, 3*(10), 1637–1661. https://doi.org/10.3390/su3101637

White, M. A. (2013). Sustainability: I know it when I see it. *Ecological Economics, 86*, 213–217. https://doi.org/10.1016/j.ecolecon.2012.12.020Williams, M. (2010). *Blind obedience*. Routledge.

Wilson, E. O. (1984). *Biophilia: The human bond with other species*. Harvard University Press.

Wilson, M. (2006). *Wandering significance: An essay on conceptual behavior*. Clarendon.

Wittgenstein, L. (1958). *Philosophical investigations* (G. E. M. Anscombe, Trans.). Macmillan.

World Commission on Environment and Development (WCED). (1987). *Our common future*. Oxford University Press.

Wu, J., & Wu, T. (2012 Sustainability indicators and indices: An overview. In C. Madu & C.-H. Kuei (Eds.), *Handbook of sustainability management* (pp. 65–86). World Scientific. https://doi.org/10.1142/9789814354820_0004

# 3 Theorizing About Sustainability

Given that sustainability is a "confusing mess of inter-related problems" (Termeer et al., 2013, p. 28), it is perhaps not surprising that it is theorized in many ways. Schlüter et al. (2022) note that proponents use disciplinary, inter-, multi-, and transdisciplinary frameworks to develop structures that explain phenomena, inform action, or improve action. They also note that many try to avoid theorizing because, among other reasons, time spent theorizing is not time spent solving urgent problems. However, just as Richard Chorley (1978) countered to his fellow geomorphologists who "instinctively reach [for their] soil augers . . . when anyone mentions theory" (p. 1), those avoiding theories are theorizing nonetheless. Theories "enter [sustainability] policy making and practice implicitly in the form of principles or insights upon which action is then based" (Schlüter et al., 2022).

Theories are important because they provide a structure for understanding, for making inferences, and for connecting those inferences into an organized whole. What do theories about the confusing messes of sustainability look like? What should they look like? Sustainability researchers and practitioners commonly invoke – explicitly or implicitly – a 'folk' view of scientific theories and the inferential practices they support, one that is often at cross purposes to the pursuit of sustainability. In 'Theory T' thinking (Wilson, 2006, 2017), the epistemological strength of theory arises from a structural base that provides strong inferential – often logical – relationships between the concepts and between the concepts and the phenomena. For science generally and sustainability in particular, this mischaracterizes theorizing and, by implication, the epistemological credibility of science (Wilson, 2006, 2017; cf. Cartwright, 2020). When used to understand sustainability, it expects structures and connections that may not be present. It elides path dependency, policy resistance, and trade-offs, sites where structures and inferential linkages are often discovered and strengthened. By encouraging widespread agreement about how to think about sustainability, it thins the notion in ways that dilute its epistemological force and normative content (cf. Miller, 2012).

Viewing sustainability theories as a rich set of interlocking practices (Cartwright, 2020; SPSP, n.d.) offers a richer way to think about sustainability

DOI: 10.4324/9781003268697-3

theorizing. I explicate this view using two interwoven theses: (1) sustainability theory as a "façade" (Wilson, 2006) of projected and extended meanings and inferences in which (2) inferences are material, i.e., based in non-exhaustive, limited-scope generalizations about the specific subject matter (J. Norton, 2003). Rather than looking for "complete exactness" (Wittgenstein, 1958, §91) or "complete clarity" (Wittgenstein, 1958, §133) from theory, the approach offers a grammar – the sometimes explicit and sometimes implicit, ever-changing rules for use (Forster, 2004) – for sustainability. It engages the contextual push and pull involved in the pursuit of sustainability. And it illuminates better the substantive and normative commitments involved that pursuit.

I argue that several representative mainstream theories of sustainability invoke Theory T thinking. I point out several limitations of this approach. I then articulate the alternative view. Theoretical approaches to sustainability such as coproduction (Arnott et al., 2020; Turnhout et al., 2020) often embody such an approach. My argument provides philosophical backing for it. And since some coproduction efforts lapse back into Theory T thinking or something close to it (Turnhout, 2020), the argument also provides some cautionary notes about how to keep from backtracking.

## Theory T

Theory T thinking involves a classical view of concepts and a separate but related view about theoretical structure and inferential practice (Wilson, 2006, 2017). I discussed the former and its limitations in Chapter 2. The latter is a thesis about how concepts are related to each other and to phenomena in ways that permit prediction and explanation. On a Theory T conception of theories, the determinate structure of a theory sanctions the inference of new theoretical relations and of the application of theoretical concepts to evidence. The power of a theory is displayed by deploying the structure and its applications across a wide range of phenomena.

One common way of expressing this is to say that a theory is a set of laws that, together with suitable conditions, allow us to deduce or otherwise strongly infer the phenomena from the behavior of underlying structures and entities. The Newtonian concept of gravitational force is commonly used as an example. This force explains why projectiles fall back to the earth (as well as many other phenomena). Given initial conditions of how fast and at what angle the projectile was thrown or fired, wind, and humidity conditions, we use the equation for the law of gravitational force to calculate the projectile's path. The laws and the conditions allow one to predict or explain by 'turning the (mathematical) crank.' Philosophical reconstructions of scientific theories, especially those that aim for logical rigor (cf. Halvorson, 2019), replace the mathematics with logic (and occasionally metamathematics). This licenses inferences *via* logical cranking.

Two consequences of this approach are important here. First, it encourages the examination of toy structures and simple forms of inferential connection. This draws attention away from the vast array of techniques that scientists use to generate connections between and among theoretical and evidential claims. Scientists use laws, but they also use idealizations, models, analogies, etc. They use several practical, observational, and manipulative techniques to say that 'this' is an instance of a theoretical concept and that 'this' thing is the same as 'that' thing. Several examples of such can be found in discussions of the generalized three-body problem (cf. Marchal, 1990), many areas of theoretical and applied physics (Cartwright, 1983; Wilson, 2006, 2017), many allele-many loci natural selection processes (Wimsatt, 1980) and economics (Cartwright, 1999). Wilson (2006) even finds them in cartography and ethno-musicology (!). Second, it encourages us to overlook the "fine-grained structure" (Wilson, 2006, p. 336) of concepts, i.e., the ways different senses are tied to different contexts. We freely talk of 'weight' as something objects like the rocks in my yard have and that astronauts in space do not. But the 'weight' being referred to is quite different. In the former, it is impressed gravitational force, the amount of effort required to move something given the environment of nearby masses. In the latter, it refers to work required to move a body relative to a local frame. By treating the two as the same, we ignore the "property dragging" (Wilson, 2006, p. 159) involved in saying both are examples of weight. We use the same word in different ways because it *seems* similar, and Theory T thinking encourages us to forget the work involved in making the connection among the different senses.

## Theory T in Sustainability Discourse

Since sustainability does not have differential equations or laws, Theory T thinking in sustainability discourse involves the more basic notion of (1) a conceptual structure or framework that (2) provides strong inferential relationships such as necessity or sufficiency among the concepts and between the concepts and the phenomena. Theories with these elements allow proponents to 'turn the crank' and understand what is and is not sustainable.

I diagnose Theory T thinking and some of its shortcomings in selected examples of sustainability theories. As examples of explanatory theories, I analyze the planetary boundaries framework and thermodynamic approaches. Matson et al.'s (2016) determinant capital assets approach; some multi-, inter-, and transdisciplinary theories; and Norton's (2005, 2015) procedural framework serve as examples of theories that inform action. I delay my discussion of theories that improve action until a later section, arguing that this style does a better job of avoiding Theory T-style thinking but that proponents still sometimes lapse into it. Given the plethora of theories in sustainability, I limit my remarks to these examples. A claim that Theory T thinking is more widespread has to be justified separately.

*Explaining*

One style of theorizing seeks to explain phenomena as logical or otherwise strong consequences of foundational concepts. This "knowledge-first" (Miller, 2012) type of theory "highlights a set of variables that have proven to be relevant for explaining a specific phenomenon, indicates high level relationships between system elements, and supports hypothesis development and testing" (Schlüter et al., 2022).[1] It is a non-mathematical analog of the Newtonian schema.

*Biophysical Explanation: Planetary Boundaries*

The planetary boundaries (PB) framework (Rockström et al., 2009; Steffen et al., 2015) influences research and policy agendas worldwide and has made its way into philosophically based general sustainability literature (Curren & Metzger, 2017). Proponents assert that sustainability will not be possible if we transgress certain pre-tipping point thresholds or "boundaries" associated with biophysical processes that regulate the stability of the earth system. Nine processes are highlighted: stratospheric ozone depletion, biodiversity loss and extinctions, chemical pollution, climate change, ocean acidification, freshwater consumption, land system change, nitrogen and phosphorus flows, and atmospheric aerosol loading. The boundaries are points beyond which, from a conservative, risk-averse point of view, it is judged impossible to maintain a "Holocene-like state" providing "a 'safe operating space' for humanity" (Steffen & Stafford Smith, 2013, p. 404). If humans are to develop and thrive, we must stay within the boundaries (Steffen et al., 2015).

The original proposal was resolutely both global and biophysical. Subsequent extensions have provided regional or local boundaries as complements for some of the processes (Steffen et al., 2015) and incorporated ethical concerns such as equity (Steffen & Stafford Smith, 2013).

There is much Theory T thinking here. While there are no laws or axiomatized postulates, the notion of a boundary does analogous work. Boundaries are "preconditions for human development" (Rockström et al., 2009) and "prerequisites for a thriving global society" (Griggs et al., 2013). Safe (and thus presumably sustainable) exists on one side; unsafe (and unsustainable) on the other. In terms of inferential strength, proponents commonly view the boundaries as allowing one to deduce or otherwise strongly infer whether activities lead to sustainability. Crossing a critical value for a threshold could, by itself, generate unacceptable change, and because the boundaries are tightly coupled transgressing one boundary poses serious risks for the others (Rockström et al., 2009). Other instances of strong inferential language are easy to find. Griggs et al. (2013) assert that the framework allows one to extract "a list of sustainability 'must-haves' for human prosperity."

The strong inference conceptions remain when the framework is expanded to include non-global and non-biophysical concerns. Considerations of scale, principles of sharing, and sustainability perspectives themselves "*jointly determine* the downscaling of the PBs – a complex process that needs to take into account the biophysical, socioeconomic, ethical and cultural dimensions" (Chen et al., 2021, emphasis added). "Combining social equity considerations with the biophysical planetary boundaries approach may therefore constitute a *necessary, and perhaps even sufficient*, condition for achieving global sustainability" (Steffen & Stafford Smith, 2013, p. 407, emphasis added).

Why be cautious about the Theory T nature of this approach? To begin, the claim that the boundaries can "determine" or "constitute a necessary, and perhaps even sufficient" condition for sustainability seems weak. The existence of boundaries (and any tipping points that lie beyond them) remains unconfirmed in many of the earth systems identified in the approach (Brook et al., 2013), particularly the nitrogen (De Vries et al., 2013) and biodiversity (Mace et al., 2014) systems. In addition, the extension to the regional scale only complements rather than specifically informs regional and local environmental efforts (Molden, 2009; Steffen et al., 2015).

Even more, the focus on thresholds (where they exist) might lead us to misidentify the necessary or sufficient (if one wants to use this inferential language) conditions for sustainability. Thresholds focus on transitions that happen relatively quickly when compared to 'normal' change. But many environmental problems do not have a baseline of normal change against which the process can be judged 'abnormal.' Consider the 'plastic soup' of particulate waste in the Pacific Ocean. This seems like a problem even though it does not seem to involve a 'tipping point' (Lewis, 2012). Likewise, the focus on thresholds neglects pressing local and regional changes that exhibit non-threshold behavior. Despite the urgency of biodiversity loss, it is not clear that this is a global threshold problem. The same holds true for some forms of chemical pollution (Cornell, 2012). Rockström et al. (2009) included these slower changes in the framework, justifying their inclusion because they contribute to the underlying resilience of the Earth system. But the choice to include them is motivated by a consideration of the resilience of the whole Earth system, not threshold behavior associated with one earth system (Cornell, 2012). More to the point, the local and regional character of these problems are likely to get lost if we focus on their global manifestations.

Plausibly, the framework's reductionistic biophysical focus makes inferences seem more direct by creating something of a toy model. Although equity concerns have been added (see below), the biophysical focus has been narrowed to the 'core' boundaries of climate change and biosphere integrity (Steffen et al., 2015). As both Hulme (2011) and Sarewitz and Pielke (2000) have argued, a biophysical frame for climate change reduces the complex relationships between the physical world and human behavior to a narrative in which the physical factor becomes the primary driver. While the planetary boundaries approach does acknowledge complex relationships among biophysical systems (Galaz et al., 2012), it pays little if any attention to interactions among coupled, self-modifying physical and human systems (Cornell, 2012). Paying more attention to these highlights the contingent connections among phenomena, weakening the focus on preconditions that jointly determine a list of 'must haves' for sustainability.

Finally, ethical criticisms of the framework indicate the contingency of the conclusions. Because the framework was constructed by university professors in the Global North, it has been criticized for lack of inclusivity in its construction and in the selection and definition of the problems (Biermann & Kim, 2020; Kim & Kotzé, 2021). For the same reason, the recent replacement of "boundaries" with "Earth systems targets" seems like "old wine in new bottles" (Biermann & Kim, 2020). The way equity concerns have been added also indicates that other inferences are possible. Steffen and Stafford Smith (2013) use a distributional notion of equity, arguing for "spatial redistributions of global phosphorus use" (and uses associated with other systems). However, justice and equity involve issues of recognition and participation as well as distribution. Who is seen and who is not seen as users who should give up and receive phosphorus allocations? Who gets to be involved in decisions about redistribution? Even more fundamentally, who gets to articulate whether phosphorus uses relate to the basic needs and functioning – the capabilities – of individuals and communities (Schlosberg 2013; Sze & London, 2008; Young, 1990)? The more one thinks about such issues, the less compelling the consequences derived from the framework seem to be.

*Biophysical Explanation: Thermodynamics*

A biophysical approach appealing to the laws of thermodynamics is intended to have similarly strong implications. Developing the framework of Georgescu-Roegen (1971), Farley and Smith (2020) argue that "in the process of producing and consuming goods . . . we draw ever nearer to the limits of the biosphere" (p. 177). The limits are determined by the first and second laws of thermodynamics. These laws are "rules which govern natural systems that fundamentally cannot be sidestepped" (p. 12). There is only one basis for sustainability: "social and economic systems are sustained *only* through environmental materials and services" (p. 180, emphasis added). "[N]atural limits to growth" are "dictated by the environment, and therefore actions in any system must adhere to the carrying capacity of the earth's natural systems" (p. 175). Echoing Tolkien, one might say there are two laws to rule them all.

On this view, much mainstream sustainability thinking gets tossed aside. Socio-ecological systems with multiple possible sustainability or resilience configurations are replaced with environmental systems characterized narrowly in thermodynamic terms. 'Confusing messes of inter-related problems' and wicked problems disappear. For Farley and Smith, talk of trade-offs – and presumably policy resistances and path dependencies – produces at best *faux* sustainability. Such talk ignores the idea that sustainability is "a process of holistic quality improvement of all systems through environmental restoration and protection" (Farley & Smith, 2020, p. 152).

This is Theory T thinking, one might say, to a T. The fundamental laws of thermodynamics provide a necessary and sufficient basis for inferring whether an action is sustainable. Proper sustainability, or "neo-sustainability," is "based on the primacy of environmental systems and the natural laws that govern those systems" (Farley & Smith, 2020, p. 184). Strong, direct inferences about actions and policies flow directly from the laws of thermodynamics. "[W]e *must* acknowledge that any production and consumption decreases the amount of usable energy and matter on earth," and "we *must* use the precautionary principle to guide our development decisions . . ." (Farley & Smith, 2020, p. 177, emphasis added). All we can do is use less matter and energy and produce as little waste as possible, with as much foresight as possible. The sense that the framework dictates the implications is even stronger when one considers that Farley and Smith state the precautionary principle as "prevention is justified if a policy or action is expected to cause harm . . . even when scientific consensus does not exist as to the nature or degree of the harm" (Farley & Smith, 2020, p. 82). By implying that precaution (of what sort?) is justified in the face of any sort of harm at any level of uncertainty and without consideration of any benefits and harms *vis-a-vis* existing practices, they employ a much-criticized strong version of the principle (Gardiner, 2006).

Why worry? As with the planetary boundaries approach, the consequences do not follow as strictly as one might think. First, the appeal to entropy in the second law of thermodynamics delivers far less than suggested. Entropy is often argued to be – at best – metaphorical or analogical (Hammond & Winnett, 2009) or – at worst – nonsense (Morowitz, 1986) when applied to far-from equilibrium systems such as the earth, where energy is being supplied constantly by the sun. This makes pronouncements about what is and is not sustainable complicated. Take subsistence agriculture. Is the total production of entropy (by animals or plants as they grow, by humans as they work the system, by the degradation of waste materials, etc.) more or less than the negentropy generated by solar input in that area? We need an analysis involving some calculations. In the absence of such, the thin, Theory T thinking covers up more than it exposes. We think we have certainty whereas in fact we just do not know whether a proposed sustainable agricultural action fits the parameters of the approach.

Second, the appeal to the second law makes everything connected since everything is energy and all actions increase entropy. But this is a thin sense of connectedness. Little is ever said about which connections matter (and which do not) from a biological, economic and/or social point of view. The resolutely biophysical framing precludes such considerations. Third, the use of entropy as a measure colors the conception of harm. Rather than engaging in a discussion of the kinds of harms, the kinds of uncertainty associated with them, or their magnitude, *anything* that degrades ecosystems is a harm (since every process produces entropy). If everything is a harm, how can any action be justified as sustainable? Fourth, the inference from the first and second laws to a carrying capacity is much too quick. There are good arguments that the biological carrying capacity for species and the territorial carrying capacity for humans must be distinguished (Aiken, 1980; cf. Attfield, 2014, p. 91). And even after they are distinguished, how carrying capacity is to be invoked needs to be specified ethically. An invocation of 'carrying capacity' may "express a desire to protect more affluent places, or a refusal to take steps to facilitate sustainable development in that country, or to assist its efforts to protect its environment" (Attfield, 2014, p. 91). In short, 'carrying capacity' and 'justice' can come unraveled. Thermodynamics is universal but not universally informative.

## Informing Action

Other 'knowledge first' approaches to sustainability focus more on informing action. Rather than characterizing "problems in terms of their causes and mechanisms *as basis for* subsequent action" (Sarewitz, quoted in Miller, 2012, p. 286, emphasis added), these theories aim to "steer complex systems towards sustainability" and "deliver desired outcomes" (Schlüter et al., 2022). Some focus on substantive matters; others emphasize procedures.

### *Informing Action Substantively: Pursuing Sustainability*

Matson et al. (2016) aim to provide a "general framework to help people interested in mobilizing knowledge to promote sustainability" (p. vii). 'Mobilizing' involves "linking knowledge to action" (Matson et al., 2016, p. 105). The "constituents of well-being" – material needs, health, education, opportunity, community, and security – have "underlying capital assets that are their determinants" (Matson et al., 2016, p. 24). The assets are natural, human, manufactured, social and knowledge capital. They are "usefully thought of as the 'state variables' of the social-environmental system. The way in which they determine, and in turn are shaped by, production and consumption processes is the crux of sustainable development" (Matson et al., 2016, p. 32).

In this framework, knowledge informs action. For Matson et al. (2016), knowledge is a fundamental productive asset. We need knowledge of the stock of capital assets, their interactions with each other, and their connections to inclusive social well-being. We also need knowledge of the dynamics of social-environmental systems because this provides a bigger, more integrated picture of what is needed. Knowledge of the dynamics "makes the invisible visible" (Matson et al., 2016, p. 63). The knowledge is necessary but not sufficient since systems are complex and interventions will have consequences beyond the intended target. Nevertheless, knowledge positions one to act.

Science helps society "to see where present trends are taking us, to discover or design new technologies and policies that might change our course, and to evaluate the possible trade-offs and implications for future generations of implementing such alternatives" (Matson et al., 2016, p. 4). To illustrate these points, Matson et al. provide case studies of London, farmer-managed irrigation systems in Nepal, fertilizer use in the Yaqui valley of Mexico, and the world's response to the ozone hole.

How is Theory T thinking manifested here? The approach assumes determinate relations among the conceptual elements and strong inferential links between knowledge and action. Capital assets are 'determinants,' i.e., factors that decisively affect the nature or outcome of well-being. Inclusive social well-being is supported by the use or consumption of goods and services "produced, drawn from and ultimately *determined by* the capital assets of the planet" (Matson et al., 2016, p. 32, emphasis added). The link between knowledge and action is similarly strong. Knowledge of capital assets allows us to "ultimately improve the ability to evaluate and track trends" (Matson et al., 2016, p. 75). Socio-environmental systems are complex, so we must be ready to adapt to new knowledge. But "better science" (Matson et al., 2016, p. 64) leads to action.

Why worry about this Theory T thinking? In terms of the conceptual basis of the proposal, the appeals to determinants and to socio-environmental system dynamics provide less clarity than supposed. Public health researchers raise questions about how determinants are framed. Despite appealing to determinants to promote a social ecological approach to health, public health continues to privilege individual- and interpersonal level measurements and interventions (Golden & Wendel, 2020). Matson, Clark, and Andersson (2016, pp. 25–29) frame their discussion largely in such terms: health is assessed using metrics of individual life expectancy, deaths of children under the age of five, and hoped for improvements in knowledge-transfer of simple hand-washing techniques and supplies. As Raphael (2006) notes, this elides questions about how the social determinants are influenced by the organization of societies, how these societies distribute resources, and how political, economic, and social forces shape organizational and distributional practices.

Questions have also been raised about the applicability of the socio-environmental systems (SES) framework. Kirchhoff et al. (2010) argue that it assumes rather than shows that the elements of the environmental system are tightly bound together, whereas other widely used ecological theories view them as far less tightly coupled. Davidson (2010) and Olsson et al. (2015) argue that the extension to loosely structured and historically dynamic social systems is problematic. Cretney (2014) argues that the framework often overlooks issues of power, agency, and inequality.

The link between knowledge and action provided by the framework is not as strong as it might appear. London is hailed as one "of the most inventive crucibles . . . in which some of the most integrated pursuits of sustainability are taking place" (Matson et al., 2016, p. 165). The bases for this claim are the passage of the Clean Air Act in 1956 and the introduction of inner-city congestion charges on automobile traffic. (Whether these policies were implemented because policy-makers were thinking about capital assets and determinants is unclear from the text.) But this is only a partial representation of well-being in London. While some advances have been made, the London Sustainable Development Commission (2017) reports that 24% of private rentals do not qualify as decent housing, housing affordability has gotten worse between 2011 and 2016, child poverty in London remains significantly higher than in the United Kingdom as a whole, fuel poverty remains at high levels (of about 10%), and close to a quarter of Londoners earned less than the London Living Wage. One wonders what the judgment might be if the discussion were widened even further beyond economic issues (of 'affordability' and 'poverty') to include, say, issues surrounding the gentrification and displacement of lower class and immigrant neighborhoods due to market forces and/or urban planning.[2]

Likewise, the elimination of ozone-damaging chlorofluorocarbons (CFCs) through the signing of the Montreal Protocol might not be all that relevant to other sustainability issues. Admittedly, this is a signature example of knowledge-based action: scientists came together to educate the public, and this was essential to mobilizing political will. However, Rayner (2006) argues that this was not a confusing mess of problems at different scales and dimensions. A small number of gases manufactured by a few companies in a few industrial countries created the problem. Moreover, substitutes were readily available, a point acknowledged by Matson et al. (2016, p. 183). A robust discussion of trade-offs due to a lack of alternatives was not needed, and lock-in due to path dependencies was not a major issue. The Protocol simply made it a certainty or near-certainty that CFCs were going to disappear from the marketplace.

### Informing Action Substantively: Multi-, Inter-, and Transdisciplinary Approaches

Multi-, inter-, and transdisciplinary approaches supposedly involve different kinds of theorizing, ones that prioritize collaboration and freedom from disciplinary constraints. However, many of these retain elements of Theory T thinking. (An exception is a certain style of transdisciplinary coproduction research, discussed later in the chapter.)

In many instances, the only difference from traditional, disciplinary theorizing is the need for integration. Multidisciplinarity "does not lead to changes in the existing disciplinary and theoretical structures" (Gibbons et al., 1994, p. 14). For example, Jerneck and Olsson (2020) marry natural and social systems to construct "a new analytical framework for integrating knowledge across the natural and social sciences" (p. 25). Likewise, interdisciplinary approaches seek "a uniform, discipline-transcending terminology or a common methodology" to create "a common framework that is shared by the disciplines involved," and transdisciplinary approaches usually rely "upon a common theoretical understanding" achieved through a "mutual interpenetration of disciplinary epistemologies" (Gibbons et al., 1994, p. 14). Wells (2013, p. 307) employs complexity theory to "provide some of the substance and blueprint for that more transdisicplinary [sic] vision" of the world. In a similar fashion, Thompson (2010) and Thompson and Norris (2021) develop a transdisciplinary approach to sustainability using systems thinking. Functional integrity, an idea associated with some sort of continuity and balance of a system, is lifted from its home base in ecology and applied broadly (Thompson, 2010, p. 220). In all three modes, unification is an achievement rather than an assumption (Klein, 2017). But the aim is still unification.

In these theoretical structures, inferences supposedly flow directly from the theoretical structure to action. Jerneck and Olssen's goal is to inform action by identifying the "leverage points" of the system. Wells (2013) uses theory similarly: approaching sustainability through the lens of complexity theory "strengthens our capacity to improve our societies and our environments" (p. 307). And although Thompson and Norris do not offer "suggestions how people should rearrange their lives or make huge changes or look for magic bullets," they do say their framework allows readers interested in action to "think critically . . . about how they can *contribute to improvements* – whether they are focused on business, or the environment, or social justice, or governance" (Thompson & Norris, 2021, p. 216, emphasis added).

In such approaches, the rhetoric of unification can all too quickly make the theory 'thin.' It assumes "that there is just one problem and that, by approaching it from many different sides, we can build up a complete picture" (Evans & Marvin, 2006, p. 1012). This ignores the fact that problem definitions themselves are based in disciplinary frameworks, framed using disciplinary and more general world view assumptions. Further, as Schmidt (2021) argues, they are often "dominated by an instrumentalist or strategic viewpoint," governed by a "collaboration and management" mindset that "develops means and instruments to reach solutions" (p. 4.). The approach becomes a call for a multitude of experts to contribute to already-defined problems and questions (Shove, 2011). This tends to avoid normative issues such as the relations of humans to nature and about the sufficiency of science in addressing the issues. In doing so, it frames the relation of the knowledge to society: it buys into "the implicit assumption [that] knowledge leads to action; *more certain* knowledge leads to *more definite* action; and *more integrated* knowledge leads to *more joined-up* action" (Hulme 2018, p. 334). In short, it reinforces a linear conception of the relation between (scientific) knowledge and action. This narrows the view of what sustainability problems are about. As West et al. (2019) argue, it excludes or sidelines the ways practitioners, policy-makers, and citizens make meaning by engaging material, social, human, and non-human actors.

Thompson (2010) and Thompson and Norris (2021) are certainly not silent on the normative aspects of sustainability. Attention to functional integrity

requires us to articulate our obligations to the institutions and natural processes that inform our identities and purposes (Thompson, 2010, p. 248). And
since sustainability raises questions about fairness between current and future people, matters of social justice are central (Thompson & Norris, 2021,
p. 131). But the strategy of using one conceptual base – functional integrity –
does thin the approach. Systems can be in equilibrium and balance in many
ways (Justus, 2021). And why be suspicious of disequilibria? Many ecologists
today "focus on a dynamic, often chaotic nature buffeted by constant distur-
bances" (Simberloff, 2014; cf. Kaus, 1992). Integrity of some sort exists in
this nature, but it is an integrity that accepts disequilibria and 'recovery' to
widely disparate states. Further, seeking integrity – and its allied behaviors
balance and stability – may lead us to quash new behaviors, especially since
integrity for some may be disintegrity for others.

### *Informing Action Procedurally: Norton*

An alternative approach to informing action involves theorizing about procedures (Miller, 2012, p. 285). The idea is to develop a participatory, nearly
always democratic process, contingent on place and time, that will lead to
consensus on action-oriented policies. Bryan Norton is a prominent proponent
of this strategy.

Norton endorses a procedural rationality that will lead to "normative sustainability." Sustainability "must be normative in the sense that it specifies, for
a given community, a sustainable path to the future, a process by which the
enduring and aspirational values important to the community are effectively
expressed and protected" (Norton, 2015, p. 102). "Procedural rationality" is
"determined by the appropriateness of the procedure employed to the problem
at hand" (Norton, 2015, p. 57) rather than some abstract ideal of logicality and
reasoning. An appropriate procedure has several elements. Appropriate procedures involve a commitment to democratic processes. The "chaotic" discussion
about sustainability must be "open to all affected parties . . . [to] ensure that the
full range of values are expressed and considered" (Norton, 2015, p. 89). The
discussion allows identification of the values that lead to sustainability.

> Normative sustainability must thus involve a careful exploration of the
> value commitments of a community and a concerted effort to determine
> which natural, physical dynamics are productive of those values and which
> support the opportunity to exercise the actions and behaviors associated
> with those values.
>
> (Norton, 2015, p. 77)

Hierarchy theory, adaptive management, and resilience are "the key concepts that lead to concern to protect key systemic features that are essential to
achieving truly strong sustainability" (Norton, 2015, p. 94). Other elements
informing the proposal include decision theory, a commitment to 'thinking

like a mountain' (in the Leopoldian sense), ecological safe minimum standards for the conservation of species, Rawlsian justice, and Habermasian discourse ethics. Overall, his proposal is that effective action moving a community toward normative sustainability can be achieved if a community articulates and adopts a set of values which it has deliberated on reflectively and democratically.

Given the focus on procedural rationality and the recognition that different communities will enact different versions of sustainability, there is decidedly less Theory T thinking here than in some of the frameworks discussed earlier. Nevertheless, a fair amount remains at the meta-level. Norton invokes several antecedent commitments that frame the discussion in ways that make inferences about actions and actions themselves follow directly from the structure. It is an open question whether the commitments are necessary or sufficient.

For one, the physical dimension of sustainability tends to dominate the framework, producing a tight connection between theory and action. The values to be protected in a community "will be associated with physical features of the environmental systems in which the community is embedded" (Norton, 2015, p. 102). What if a community decides to privilege social and cultural concerns for a particular issue? The attitude toward the "key concepts" of hierarchy theory, adaptive management and resilience imposes similarly tight connections. For Norton, "the space-time boundaries of an environmental problem *must* be understood within a hierarchical systems theoretic perspective" (Zia, 2018, p. 170, emphasis added). Hierarchy theory includes antecedent commitments to viewing systems as operating on different spatiotemporal scales, components of systems being a part of larger parts while simultaneously being made of smaller parts, and the near-decomposability a system under a particular mode of observation. Must all sustainability problems necessarily be approached through such lenses? Norton (2015) claims that commitments to the framework's elements are ones a community "*would* espouse as the outcome of an appropriate process of deliberation and social learning" (p. 78, emphasis added). Norton (2015, pp. 218–257) does discuss at length two case studies to show that communities can enact stronger sustainability if the procedures *are* embraced, but the source of confidence in the claim that such processes *necessarily would* lead to them is puzzling.

Specific commitments to the processes of deliberation and social learning also produce stronger inferential links between theory and action than might exist in many cases. Many Native American approaches to sustainability involve deep relational conceptions of nature and society over several generations. Communities are deeply committed to preserving relations and natural bases for well-being. Yet committees of elites are often responsible for articulating and governing such commitments (Burkhart, 2019; Whyte et al., 2018). What Norton calls strong sustainability seems possible without using democratic processes. Additionally, Norton (2015) – and in a strikingly similar vein Mitchell (2007) – narrows the discussion to communities that are willing to bridge differences and to frame disagreements in ways that encourage working together to find viable ways to coexist. Not all communities do, will or even could proceed in this manner. Many might wish to not work together as one, considering a creative balancing of values in tension with each other to be a virtue. Even if the members consider themselves to belong to the same community, they may continue to espouse different sustainability values (Piso et al., 2016; Zia, 2018). Social decision-making processes designed to elicit agreement may summon *more* diverse and conflicting ideas (Brister, 2018). If one drops the assumption about being willing to bridge differences, it becomes a live question whether the community can even agree on a set of values and procedural rules that might lead to support of the values (Hirsch, 2018). Adding questions about the distorting influence of power relations (economic, political, etc.) and institutions (at many scales and levels) only circumscribes the link between theory and action more narrowly (cf. Zia, 2018).

As mentioned above, Norton discusses several successes. So the theory does work under particular conditions. But it is not obvious that the framework can link knowledge – of procedures and of particular theoretical commitments – to action more generally. A 'toy structure' seems to have been created.

## Sustainability Theorizing as Patchwork Façades of Material Inferences

Theory T thinking does have its place, as when it solves a problem by proving the mathematical self-consistency of a formalism (Wilson, 2017, pp. 151–152). But for more common, garden-variety theorizings – ones that might help us understand the confusing mess of problems that define sustainability – it hides the creative work involved in linking concepts and applying them to diverse situations in the world. That is, it hides the ways in which sustainability comes to have meaning in specific places and times. To see this work, it is helpful to view a theory as a façade of projected and extended meanings (Wilson, 2006, 2017) linked by material inference (J. Norton, 2003, 2021; cf. Wilson, 2017; Love, 2012).

Wilson uses the general notion of a façade or atlas (he uses the terms interchangeably) to characterize the relations that connect concepts. A façade is an organized structure of projected and extended meanings of individual concepts or sets of concepts. Interactions with various kinds of material stuff produce different senses of the same concept, and the structure in the façade lays out the links among them. As an example, consider the characterization of classical physics as billiard ball mechanics involving collisions (Wilson, 2006, 180ff.). While texts standardly treat the balls as rigid bodies when they collide with each other, experience tells us that most real spheres readily distort under impact. Dribble a basketball or watch a baseball being hit in slow motion. For such collisions, the equations of rigid body mechanics are modified, often with entirely different mathematics. Different kinds of impacts lead to yet more different descriptive and mathematical treatments. Examples include collisions that generate wave movements (think of the waves created when your finger 'collides' with a guitar string) and high speed collisions at explosive velocities (go on YouTube and watch videos of bullets hitting – well – just about anything). The overall result is different descriptive and theoretical patches associated with different kinds of materials and behaviors. Since we most often encounter situations where the rigid body approximation is good enough for our purposes, we tend to forget or ignore these entirely different senses of 'collision.'

For Wilson, patches occur because of who we are and what the world is like. To paraphrase Winnie the Pooh, we are beings of little brain. "We are not supernatural intellects; we . . . must cobble together and redirect our modest computational inheritance in the pursuit of more sophisticated objectives" (Wilson, 2017, p. 4). Wimsatt (2007) makes a similar argument that we must pay attention to our error-prone, cognitively limited minds if we are to understand how science works. Both Wilson and Wimsatt stress that systems in the world are messy: "nature rarely arranges its affairs for our calculational convenience" (Wilson, 2017, p. 364).

Even if one does not accept these reasons why the patches exist, they do exist (once one looks for them). Sometimes the connections between patches are smooth. The example of 'weight,' mentioned earlier, is an example: the behavior of the 'weightless' astronaut meets our expectations of what we should see if all impressed gravitational force were reduced to zero (Wilson, 2006, pp. 328–335). At other times the connections are discontinuous and irregular. Take 'hardness.' We evaluate the hardness of materials by applying tests: we squeeze the object, hit it with a hammer, try to scratch it, see whether it is resistant to wear, etc. Each of these tests probes a different property.

> To the metallurgist, hardness is the resistance to indentation; to the design engineer, a measure of flow stress; the lubrication engineer, the resistance to wear; to the mineralogist, the resistance to scratching; and to the machinist, the resistance to cutting.
>
> (Fee, Segabache & Tobolski, 1985; as cited in Wilson, 2006, p. 337)

Tests of one property do not easily coincide with hardness tests developed for others. Testing gemstones for resistance to indentation is, in most circumstances, not advisable. Few tests of the hardness of pearls or rubies aim to dent or, worse, smash them.

The lack of transfer is why patches are needed. Our probings (scratching, tapping, hitting, etc.) are relatively portable and serve to orient us across the different senses. I *could* hit the gem. But this does not give me a good sense of, much less data about, its hardness in relation to other gems (since 'breakability under the same impact' is not something about gems that we are usually interested in, even if we are willing to sacrifice the gem). The everyday claim that 'X is harder than Y' grounds a particular claim. When needed, patches to other senses are constructed. As we make the comparative judgment, we shift meanings and contexts all the time. "Our abilities to shift contexts quickly along with the rarity with which we compare discordant patches provides an adequate data control to keep our usage of 'is hard' from collapsing into deductive incoherence" (Wilson, 2006, p. 346). We manage in practice. But it is only by attending to the fine structure that we uncover the warrants for extensions of previous practices into new territories.

In the patchwork façades, inferences are licensed materially. Material inference is a local, non-formal theory of induction powered by background knowledge. Rather than inferences being licensed by form, as in *modus ponens* and *modus tollens*, inferences are licensed by facts relevant to the inference. Contrast the following (Norton, 2003, p. 649):

P1) Some samples of the element bismuth melt at 271 degrees C.
C1) Therefore, all samples of the element bismuth melt at 271 degrees C.
P2) Some samples of wax melt at 91 degrees C.
C2) Therefore, all samples of wax melt at 91 degrees C.

The first inference is inductively strong. The second is weak. Why?

> We are licensed to infer from the melting point of some samples of an element to the melting point of all samples by a fact about elements: their samples are generally uniform in their physical properties. So if we know the physical properties of one sample of the element, we have a license to infer that other samples will most likely have the same properties. The license does not come from the form of the inference . . . It comes from a fact relevant to the material of the induction. There are no corresponding facts for the induction on wax, so the formal similarity between the two inductions is a distraction.
>
> (Norton, 2003, p. 650)

The "material postulates," or "facts," involve background or contextual knowledge (of the way the materials behave, gathered empirically and inductively) that is not represented explicitly in the inference. Love (2012) offers an instructive example of material inferences in evolutionary developmental biology (EvoDevo). The various principles, rules and hypotheses used to explain the forms and structures of organisms all involve defeasible generalizations generated from empirical inquiry.

For sustainability, several points are important. First, the difference in the strength of the warrants comes from "material postulates" about the stuffs of bismuth and wax. Given the wicked messes of sustainability problems, the warrants backing specific inferences are not obvious. Discovering them is crucial. Are we dealing with facts that are more like bismuth or more like wax? Reliability in a particular mess seems far more relevant than the 'truth' (in any transcendental sense) of the facts. Second, the notion of induction here is broad, encompassing analogical inference, abduction, projection, etc. (Shech & Parker, 2021). Wilson (2017) analyzes many inference structures used to apply classical mechanics to extended matter. Without addressing this issue directly, I believe the discussion that follows indicates that just as many are used in sustainability discourse. Third, while Norton focuses on physical facts, licenses for inferences can come from different kinds of uniformities. In sustainability discussions, values are often taken to be secured by uniformity (of whatever kind) in experience (of whatever kind). Last, understanding the correctness of the inference is a kind of "practical mastery of a certain kind of inferentially articulated doing: responding differentially according to the circumstances of proper application of a concept, and distinguishing the proper inferential consequences of such application" (Brandom, 2000, pp. 63–64). Since we do not know already what sustainability will look like in any given situation, we learn by doing. We might like to lay down the law about sustainability, but we should recognize that "it is impossible to move to the desired state in a straight line since there are too many variables and uncertainties" (Kemp et al., 2007, p. 11).

What does sustainability theory look like when viewed as a patchwork of material inferences? Consider the many different interpretations of what is to be sustained, what is to be developed and what the relation between the two should be outlined in *Our Common Future* (NRC, 1999). Each interpretation is linked to a set of material postulates that provides a basis for asserting a claim of sustainability. As new claims are made, the meaning is dragged over to the new context and changed in the process. Rather than strong inferences derived from an *a priori* collection of principles, inferences are supported (or not) by the strength of particular local connections. For instance, deforestation in the Amazon might be a major cause for concern not because it is an example of land use change generally – as in the planetary boundaries approach – but "because of the interplay of that particular patch of vegetation with the processes influencing global water and energy balance" (Cornell, 2012). Likewise, the human perturbation of biogeochemical flows of essential nutrient elements "matters more in places where ecosystem function is jeopardized . . . and in some places, these changes in function have global consequences" (Cornell, 2012).

As people make new claims about the local regularities underpinning their conception of sustainability, they are challenged. The disagreements provide opportunities to see the shape and content of specific connections. For example, Whyte et al. (2018) push back against the common strategy of recognizing Indigenous people's cultures as holding general insights about sustainability. In order to understand sustainability in the Menominee Nation in particular (and, by extension, other Indigenous peoples living in settler states), they note that sustainability must be specified in the context of settler colonialism. Today, Menominee Tribal Enterprises oversees forest management and sawmill operations in what is historically the first practice of sustainable forestry in the United States. In this (and other Indigenous collectives), planning involves ceremonies that "express hope and emotional interpretations of the future" (Whyte et al., 2018, p. 156), researching knowledge archives, and other techniques for forecasting future scenarios. In all of the activities, "Indigenous peoples imagine themselves strategically in ways that are not reliant on settler and other oppressive desires, discourses, and needs. ... The forest relies on Menominee history, culture, and knowledge in resistance to settle colonial oppression" (Whyte et al., 2018, p. 164). The Menominee create a patch from a conception of sustainability parading as universal to their own conception, one that involves the facts of settler colonialism. They re-represent the 'what' of a supposed general sustainability into something that cannot be 'sustainability for all.' The facts relevant to sustainability in that situation provide the Menominee with a license to infer 'this is sustainability for us.' The contestation allows us to see how the transition from one sense to the other is made. The meanings are not translated without loss. Each is a material inference tied to specific assumptions about what and who sustainability is for.

While this is a single case, other examples support the larger claim that meaning making about sustainability is made locally with bounded 'material postulates,' i.e., 'facts' (cf. Sze, 2018). I have used the notion of a 'patchwork' façade, but other type of facades are equally possible. The important point for my purposes is that asking for and talking about examples is crucial. Whether a process or procedure should count as 'sustainable' comes from "a grasp on things which, although quite unarticulated, may allow us to formulate reasons and explanations when challenged" (Taylor, 1999, p. 32). Seeing and hearing the reasons in a challenge helps articulate the inferential connections. The general overall pattern of reasoning involves taking a practice as exemplary (even if problematic), asking about its connections to goals and values, settling on some relevant 'material postulates,' and then extending that sense to a new practice or site. Just as in physics, 'property dragging' will occur. Viewed from far away, the senses may not appear to have anything to do with each other. Closer, we can see a kind of natural normativity arising from the back and forth questioning.

Within sustainability theorizing, this account of theory structure and inferential practices most closely resembles certain forms of participatory research. This style of research aims to improve action. Theories "emerge through knowledge co-production and transdisciplinary methodologies . . . aiming to empower and enable actors to create change on their own behalf" (Schlüter et al., 2022). This style includes a wide variety of names and practices, including coproduction, transdisciplinarity, science-policy interface studies, democratization of expertise, and knowledge brokering (Turnhout et al., 2020). To simplify the discussion, I focus on coproduction. Over the past decade, knowledge coproduction has shifted to the mainstream of scientific practice, and the strategy is now commonplace in sustainability research even though what it means is highly variable (Norström et al., 2020; West et al., 2019). The philosophical rationale I have developed can usefully be thought of as supporting coproduction as a general framework and, in addition, as supporting particular interpretations of coproduction.

Coproduction involves diverse types of expertise, knowledge, and actors in the production of context-specific knowledge and pathways toward sustainability. Such research is context-based, pluralistic, goal-oriented and interactive (Norström et al., 2020). Coproduction emphasizes the experiential, collaborative and partial nature of all attempts at addressing a problem. A key premise is that scientific expertise alone is not sufficient (Turnhout et al., 2020, p. 16). Scholars work together with diverse social actors to produce – not just inform – change. Like all other actors, researchers are "embedded" in the system. All proceed inclusively from the bottom up. They do not impose a particular conceptual structure on to the phenomena or presume that any given stakeholder knows what 'the problem' is. What the problem is, who it is a problem for, who can address the problem, how they address it – all vary from project to project.

Theorizing here happens within practice, within the process of producing change. Knowledge is "not applied to action, but rather produced and used within a situation, shaped by the outcomes that emerge during the process" (Schlüter et al., 2022). The understanding involves coming to know the physical, social, cultural, and institutional factors affecting the situation and becoming aware of the ways in which one's own framing of the issue shapes the conception of the factors and their effects. A patch about what sustainability means in that context is constructed. From there, 'patching' to new situations is possible (if that is helpful). As a collection of examples is generated, a grammar – the acceptable or unacceptable rules of use – of sustainability emerges.

In attending to context, pluralism, goals, and interactiveness, one must be careful not to fall back into a structuralist, Theory T-style construal where legitimacy is conferred by adoption of a particular theoretical base and the inferences it licenses. Schlüter et al. (2022) are a cautionary case in point. They embrace a diversity of theorizing practices due to a prior commitment to the theory of socio-ecological systems (SES). Theories in sustainability science "are informed by and reflect the diversity of world-views, values and goals that characterize SES." Pluralism about theorizing is needed due to "the complexity of SES." Worries voiced earlier about the extension of SES to social systems are relevant here. If SES is not as applicable as envisaged, the embrace of pluralism arises out of a mistaken commitment to a framework. Additionally, Schlüter et al. do not seem to consider embedded research in a non-SES framework. Nor do they consider a pluralistic approach to theorizing that is not based in such. Practice theorists almost universally endorse open-ended and collaborative learning-by-doing among all members of the affected community (West et al., 2019), yet they do not feel the need to adhere strongly to SES.

Other factors can also pull coproduction back into Theory T thinking. As Turnhout et al. (2020, p. 15) note, much literature on coproduction is programmatic or methodological, "present[ing] best practices and lessons learnt, and offer[ing] checklists of factors and conditions for success." This lets researchers pull away from the messy, power- and politically laden contexts of actual coproduction processes. Additionally, some of the thinning may come from seeing coproduction as "shorthand for participatory modes of knowledge" (Turnhout et al., 2020, p. 15). As noted earlier regarding interdisciplinary research, this frames the issue as an instrumental question governed by a 'collaboration and management' mindset. One can focus on the methodological questions involved in getting people to talk to and with each other, not addressing the ways in which "the ways in which we know and represent the world (both nature and society) are inseparable from the ways in which we choose to live in it" (Jasanoff, 2004, as cited in Turnhout et al., 2020, p. 15).

As my advocacy of a particular style of theorizing would indicate, I do not think theory should be avoided. My argument only places bounds on kinds of theorizing that are likely to be useful for sustainability. Postulating universal substantive and/or procedural foundations should be avoided because this strategy does not address how meanings are made and inferential connections forged. As the connections are forged, material postulates of various strengths emerge. This allows us to locate the local meanings of sustainability more successfully (cf. Chapter 2), understand how the facts relate to the evidence of sustainability (cf. Chapter 4) and navigate the responsibilities involved (cf. Chapter 5). Overall, a robust pluralism about theories of sustainability – and about sustainability itself – seems necessary to generate those meanings.

## Notes

1   Since the relationship between sustainability and resilience is complicated, with some seeing them as linked and others as separate objectives (Marchese et al., 2018), this chapter discusses the complex socio-ecological systems (SES) framework only in the context of theories that employ it. I leave for another time questions of whether Theory T thinking is present in the SES framework itself and/or in approaches that link sustainability and resilience.
2   Thanks to my colleague Efadul Huq for calling my attention to these issues.

## References

Aiken, W. (1980). The "carrying capacity" equivocation. *Social Theory and Practice*, *6*(1), 1–11. https://doi.org/10.5840/soctheorpract19806114

Arnott, J. C., Mach, K. J., & Wong-Parodi, G. (2020). Editorial overview: The science of actionable knowledge. *Current Opinion in Environmental Sustainability*, *42*, A1–A5. https://doi.org/10.1016/j.cosust.2020.03.007

Attfield, R. (2014). *Environmental Ethics,* 2nd ed. Polity Press.

Biermann, F., & Kim, R. E. (2020). The boundaries of the planetary boundary framework: A critical appraisal of approaches to define a "safe operating space" for humanity. *Annual Review of Environment and Resources, 45*(1), 497–521. https://doi.org/10.1146/annurev-environ-012320-080337

Brandom, R. (2000). *Articulating reasons.* Harvard University Press.

Brister, E. (2018). Proceduralism and expertise in local environmental decision-making. In S. Sarkar & B. Minteer (Eds.), *A sustainable philosophy – The work of Bryan Norton* (pp. 151–166). Springer.

Brook, B. W., Ellis, E. C., Perring, M. P., Mackay, A. W., & Blomqvist, L. (2013). Does the terrestrial biosphere have planetary tipping points? *Trends in Ecology & Evolution, 28*(7), 396–401. https://doi.org/10.1016/j.tree.2013.01.016

Burkhart, B. (2019). *Indigenizing philosophy through the land: A trickster methodology for decolonizing environmental ethics and indigenous futures.* Michigan State University Press.

Cartwright, N. (1983). *How the laws of physics lie.* Oxford University Press.

Cartwright, N. (1999). *The dappled world: A study of the boundaries of science.* Cambridge University Press.

Cartwright, N. (2020). Middle-range theory. *Theoria: An International Journal for Theory, History and Foundations of Science, 35*(3), 269–323. https://doi.org/10.1387/theoria.21479

Chen, X., Li, C., Li, M., & Fang, K. (2021). Revisiting the application and methodological extensions of the planetary boundaries for Sustainability Assessment. *Science of the Total Environment, 788*, 147886. https://doi.org/10.1016/j.scitotenv.2021.147886

Chorley, R. (1978). Bases for theory in geomorphology. In C. Embleton, D. Brunsden & D. Jones (Eds.), *Geomorphology: Present problems and future prospects* (pp. 1–13). Oxford University Press.

Cornell, S. (2012). On the system properties of the planetary boundaries. *Ecology and Society, 17*(1): r2. https://doi.org/10.5751/es-04731-1701r02

Cretney, R. (2014). Resilience for whom? Emerging critical geographies of socio-ecological resilience. *Geography Compass, 8*(9), 627–640. https://doi.org/10.1111/gec3.12154

Curren, R., & Metzger, E. (2017). *Living well now and in the future.* MIT Press.

Davidson, D. J. (2010). The applicability of the concept of resilience to social systems: Some sources of optimism and nagging doubts. *Society & Natural Resources, 23*(12), 1135–1149. https://doi.org/10.1080/08941921003652940

de Vries, W., Kros, J., Kroeze, C., & Seitzinger, S. P. (2013). Assessing planetary and regional nitrogen boundaries related to food security and adverse environmental impacts. *Current Opinion in Environmental Sustainability, 5*(3–4), 392–402. https://doi.org/10.1016/j.cosust.2013.07.004

Evans, R., & Marvin, S. (2006). Researching the sustainable city: Three modes of interdisciplinarity. *Environment and Planning A: Economy and Space, 38*(6), 1009–1028. https://doi.org/10.1068/a37317

Farley, H., & Smith, Z. (2020). *Sustainability: If it's everything, is it nothing?* 2nd ed. Routledge.

Forster, M. (2004). *Wittgenstein on the arbitrariness of grammar.* Princeton University Press.

Galaz, V., Crona, B., Österblom, H., Olsson, P., & Folke, C. (2012). Polycentric systems and interacting planetary boundaries — Emerging governance of climate change–ocean acidification–Marine biodiversity. *Ecological Economics, 81*, 21–32. https://doi.org/10.1016/j.ecolecon.2011.11.012

Gardiner, S. M. (2006). A core precautionary principle. *Journal of Political Philosophy, 14*(1), 33–60. https://doi.org/10.1111/j.1467-9760.2006.00237.x

Georgescu-Roegen, N. (1971) *The entropy law and the economic process.* Harvard University Press.

Gibbons, M., Limoges, C., Nowotny, H., Schwartzman, S., Scott, P., & Trow, M. (1994). *The new production of knowledge.* Sage.

Golden, T. L., & Wendel, M. L. (2020). Public health's next step in advancing equity: Re-evaluating epistemological assumptions to move social determinants from theory to practice. *Frontiers in Public Health, 8*, 1–7. https://doi.org/10.3389/fpubh.2020.00131

Griggs, D., Stafford-Smith, M., Gaffney, O. *et al.* (2013). Sustainable development goals for people and planet. *Nature, 495*, 305–307. https://doi.org/10.1038/495305a

Halvorson, H. (2019). *The logic in philosophy of science.* Cambridge University Press.

Hammond, G., & Winnett, A. (2009). The influence of thermodynamic ideas on ecological economics: An interdisciplinary critique. *Sustainability, 1*(4), 1195–1225. https://doi.org/10.3390/su1041195

Hirsch, P. (2018). Sustainable change in a fractured world. In S. Sarkar and B. Minteer (Eds.), *A sustainable philosophy – The work of Bryan Norton* (pp. 97–98). Springer.

Hulme, M. (2011). Reducing the future to climate: A story of climate determinism and reductionism. *Osiris, 26*(1), 245–266. https://doi.org/10.1086/661274

Hulme, M. (2018). 'Gaps' in climate change knowledge: Do they exist? can they be filled? *Environmental Humanities, 10*(1), 330–337. https://doi.org/10.1215/22011919-4385599

Jerneck, A., & Olsson, L. (2020). Theoretical and methodological pluralism in sustainability science. In T. Mino & S. Kudo (Eds.), *Framing in sustainability science: Theoretical and practical approaches* (pp. 17–33). Springer Nature.

Justus, J. (2021). *The philosophy of ecology: An introduction.* Cambridge University Press.

Kaus, A. (1992). Taming the wilderness myth. *BioScience, 42*(4), 271–279. https://doi.org/10.2307/1311675

Kemp, R., Loorbach, D., & Rotmans, J. (2007). Transition management as a model for managing processes of co-evolution towards sustainable development. *International Journal of Sustainable Development & World Ecology, 14*(1), 78–91. https://doi.org/10.1080/13504500709469709

Kim, R. E., & Kotzé, J. L. (2021). Governing the complexity of planetary boundaries: A state-of-the-art analysis of social science scholarship. In D. French & J.L. Kotzé (Eds.), *Research handbook on law, governance and planetary boundaries* (pp. 45–64). Edward Elgar Publishing. https://doi.org/10.4337/9781789902747.00010

Kirchhoff, T., Brand, F. S., Hoheisel, D., & Grimm, V. (2010). The one-sidedness and cultural bias of the resilience approach. *GAIA - Ecological Perspectives for Science and Society, 19*(1), 25–32. https://doi.org/10.14512/gaia.19.1.6

Klein, J. (2017). Transdisciplinarity and sustainability: Patterns of definition. In D. Fam, J. Palmer, C. Riedy & C. Mitchell (Eds.), *Transdisciplinary research and practice for sustainability outcomes* (pp. 7–21). Routledge.

Lewis, S. (2012). We must set planetary boundaries wisely. *Nature, 485*, 417. https://doi-org/10.1038/485417a

London Sustainable Development Commission (2017). *London's quality of life indicators report 2017*. https://data.london.gov.uk/londons-quality-of-life-indicators-report/.

Love, A. C. (2012). Theory is as theory does: Scientific practice and theory structure in biology. *Biological Theory, 7*(4), 325–337. https://doi.org/10.1007/s13752-012-0046-2

Mace, G. M., Reyers, B., Alkemade, R., Biggs, R., Chapin, F. S., Cornell, S. E., Díaz, S., Jennings, S., Leadley, P., Mumby, P. J., Purvis, A., Scholes, R. J., Seddon, A. W. R., Solan, M., Steffen, W., & Woodward, G. (2014). Approaches to defining a planetary boundary for biodiversity. *Global Environmental Change, 28*, 289–297. https://doi.org/10.1016/j.gloenvcha.2014.07.009

Marchal, C. (1990). *The three-body problem*. Elsevier.

Marchese, D., Reynolds, E., Bates, M. E., Morgan, H., Clark, S. S., & Linkov, I. (2018). Resilience and sustainability: Similarities and differences in environmental management applications. *Science of the Total Environment, 613–614*, 1275–1283. https://doi.org/10.1016/j.scitotenv.2017.09.086

Matson, P., Clark, W., & Andersson, K. (2016). *Pursuing sustainability: A guide to the science and practice*. Princeton University Press.

Miller, T. R. (2012). Constructing sustainability science: Emerging perspectives and research trajectories. *Sustainability Science, 8*(2), 279–293. https://doi.org/10.1007/s11625-012-0180-6

Miller, T. R. (2015). *Reconstructing sustainability science*. Routledge.

Mitchell, S. D. (2007). The import of uncertainty. *The Pluralist, 2*(1), 58–71. https://doi.org/10.2307/20708888

Molden, D. (2009). Planetary boundaries: The devil is in the detail. *Nature Climate Change, 1*(910), 116–117. https://doi.org/10.1038/climate.2009.97

Morowitz, H. (1986). Entropy and nonsense. *Biology and Philosophy, 1*, 473–476. https://doi.org/ 10.1007/bf00140964

National Research Council (NRC) (1999). *Our common journey: A transition toward sustainability*. National Academies Press.

Norström, A. V., Cvitanovic, C., Löf, M. F., West, S., Wyborn, C., Balvanera, P., ... & Österblom, H. (2020). Principles for knowledge co-production in sustainability research. *Nature Sustainability, 3*(3), 182–190. https://doi.org/10.1038/s41893-019-0448-2

Norton, B. (2005). *Sustainability: A philosophy of adaptive ecosystem management*. University of Chicago Press.

Norton, B. (2015). *Sustainable values, sustainable change: A guide to environmental decision making*. University of Chicago Press.

Norton, J. (2003). A material theory of induction. *Philosophy of Science, 70*, 647–670. https://doi.org/10.1086/378858

Norton, J. (2021). *The material theory of induction*. University of Calgary Press.

Olsson, L., Jerneck, A., Thoren, H., Persson, J., & O'Byrne, D. (2015). Why resilience is unappealing to social science: Theoretical and empirical investigations of the scientific use of resilience. *Science Advances, 1*(4), e1400217. https://doi.org/10.1126/sciadv.1400217

Piso, Z., Werkheiser, I., Noll, S., & Leshko, C. (2016). Sustainability of what? Recognising the diverse values that sustainable agriculture works to sustain. *Environmental Values, 25*(2), 195–214. https://doi.org/10.3197/096327116x14552114338864

Raphael, D. (2006). Social determinants of health: Present status, unanswered questions, and future directions. *International Journal of Health Services, 36*(4), 651–677. https://doi.org/10.2190/3mw4-1ek3-dgrq-2crf

Rayner, S. (2006). Wicked problems: Clumsy solutions–diagnoses and prescriptions for environmental ills. *Jack Beale Memorial Lecture on Global Environment*. https://core.ac.uk/download/pdf/288283455.pdf

Rockström, J., Steffen, W., Noone, K., Persson, Å., Chapin III, F. S., Lambin, E., . . . & Foley, J. (2009). Planetary boundaries: Exploring the safe operating space for humanity. *Ecology and Society*, *14*(2), 32. http://www.ecologyandsociety.org/vol14/iss2/art32/

Sarewitz, D., & Pielke Jr., R. (2000). Breaking the global-warming gridlock. *The Atlantic Monthly*, *286*, 55–64.

Schlosberg, D. (2013). Theorizing environmental justice. *Environmental Politics*, 22, 37–55. http://dx.doi.org/10.1080/09644016.2013.755387

Schlüter, M., Caniglia, G., Orach, K., Bodin, Ö., Magliocca, N., Meyfroidt, P., & Reyers, B. (2022). Why care about theories? Innovative ways of theorizing in sustainability science. *Current Opinion in Environmental Sustainability*, *54*, 101154. https://doi.org/10.1016/j.cosust.2022.101154

Schmidt, J. (2021). *Philosophy of interdisciplinarity*. Routledge.

Shech, E., & Parker, W. (2021). Introduction to Norton's material theory. *Studies in History and Philosophy of Science Part A*, *85*, 30–33. https://doi.org/10.1016/j.shpsa.2020.12.003

Shove, E. (2011). On the difference between chalk and cheese—a response to whitmarsh et al's comments on "Beyond the abc: Climate change policy and theories of social change." *Environment and Planning A: Economy and Space*, *43*(2), 262–264. https://doi.org/10.1068/a43484

Simberloff, D. (2014). The "Balance of nature"—evolution of a panchreston. *PLoS Biology*, *12*(10), e1001963. https://doi.org/10.1371/journal.pbio.1001963

SPSP (Society for the Philosophy of Science in Practice) (n.d.). *Mission statement*. https://philosophy-science-practice.org/about/mission-statement.

Steffen, W., Richardson, K., Rockström, J., Cornell, S. E., Fetzer, I., Bennett, E. M., Biggs, R., Carpenter, S. R., de Vries, W., de Wit, C. A., Folke, C., Gerten, D., Heinke, J., Mace, G. M., Persson, L. M., Ramanathan, V., Reyers, B., & Sörlin, S. (2015). Planetary boundaries: Guiding human development on a changing planet. *Science*, *347*(6223), 1259855. https://doi.org/10.1126/science.1259855

Steffen, W., & Stafford Smith, M. (2013). Planetary boundaries, equity and global sustainability: Why wealthy countries could benefit from more equity. *Current Opinion in Environmental Sustainability*, *5*(3–4), 403–408. https://doi.org/10.1016/j.cosust.2013.04.007

Sze, J. (Ed.) (2018). *Sustainability: Approaches to environmental justice and social power*. NYU Press.

Sze, J., & London, J. K. (2008). Environmental justice at the crossroads. *Sociology Compass*, *2*(4), 1331–1354. https://doi.org/10.1111/j.1751-9020.2008.00131.x

Taylor, C. (1999). To follow a rule. In R. Schusterman (Ed.), *Bourdieu: A critical reader* (pp. 29–44). Blackwell.

Termeer, C., Dewulf, A., & Breeman, G. (2013). Governance of wicked climate adaption problems. In J. Kneiling & W. Leal Filho (Eds.), *Climate change governance* (pp. 27–39). Springer-Verlag.

Thompson, P. (2010). *The agrarian vision*. University Press of Kentucky.

Thompson, P., & Norris, P. (2021). *Sustainability: What everyone needs to know*. Oxford University Press.

Turnhout, E., Metze, T., Wyborn, C., Klenk, N., & Louder, E. (2020). The politics of co-production: Participation, power, and transformation. *Current Opinion in Environmental Sustainability, 42*, 15–21. https://doi.org/10.1016/j.cosust.2019.11.009

Wells, J. (2013). *Complexity and sustainability*. Routledge.

West, S., van Kerkhoff, L., & Wagenaar, H. (2019). Beyond "linking knowledge and action": Towards a practice-based approach to transdisciplinary sustainability interventions. *Policy Studies, 40*(5), 534–555. https://doi.org/10.1080/01442872.2019.1618810

Whyte, K., Caldwell, C., & Schaefer, M. (2018). Indigenous lessons about sustainability are not just for "all humanity. In J. Sze (Ed.), *Sustainability: Approaches to environmental justice and social power* (pp. 149–179). New York University Press.

Wimsatt, W. (1980). Reductionistic research strategies and their biases in the units of selection controversy. In T. Nickles (Ed.), *Scientific discovery, Volume II: Historical and scientific case studies* (pp. 213–259). Reidel.

Wilson, M. (2006). *Wandering significance: An essay on conceptual behavior*. Oxford University Press.

Wilson, M. (2017). *Physics avoidance: And other essays in conceptual strategy*. Oxford University Press.

Wimsatt, W. C. (2007). *Re-engineering philosophy for limited beings: Piecewise approximations to reality*. Harvard University Press.

Wittgenstein, L. (1958). *Philosophical investigations* (G. E. M. Anscombe, Trans.). Macmillan.

Young, I. (1990). *Justice and the politics of difference*. Princeton University Press.

Zia, A. (2018). Adaptive management in social ecological systems – Taming the wicked? In S. Sarkar & B. Minteer (Eds.), *A sustainable philosophy – The work of Bryan Norton* (pp. 167–187). Springer International.

# 4 Evidence for Sustainability

In the complex, wicked problems of sustainability, it is seldom clear whether our actions lead to sustainability. Actions may have consequences at remote times, places, and scales. Actions we expect to have consequences may have none or the wrong ones. We need evidence to know whether we are becoming more sustainable. While many types of evidence are used in sustainability discourse, perhaps the most prominent are indicators and amalgams of such known as indices. The 1992 Rio Summit call for sets of indicators to monitor progress toward sustainability was wildly successful. An "indicator industry" (Sébastien & Bauler, 2013) has produced an "indicator zoo" (Pintér et al., 2012) at global, national, and local scales.

What kind of evidence are sustainability indicators? Philosophers, historians, and sociologists of science have not paid much attention to this question.[1] In much mainstream sustainability discourse, indicators and the roles they play reflect a philosophical approach to data[2] that Lloyd (2012) has dubbed "direct empiricism." Indicators are seen as hard, objective evidence that can inform and steer policy. This thin interpretation strips indicators of context and, in so doing, raises several worries. In contrast, an 'indicators as models' conception connects indicators to their contexts and the contestations over their construction and use. This "complex empiricist" (Lloyd, 2012) approach asserts that data and evidence are inevitably laden with assumptions and theory. This offers a more robust understanding of the construction and use of indicators. In addition, it shows how indicators relate to the phenomena of trade-offs, path dependencies, and policy resistance. Finally, it highlights the multiple normative dimensions of data, allowing us to see indicators as artifacts that have political and moral implications.

Several authors in the sustainability literature have argued for what I am calling a model-based approach (Gallopin, 2018; Hansson et al., 2019; Lehtonen et al., 2016; Merino-Saum et al., 2020; Sari et al., 2018). Bell and Morse (2018), leading scholars on sustainability indicators, note however that "walking the talk" of indicators as contested and contextual remains uncommon. A complex empiricist approach to indicators adds more reasons to interpret indicators thickly as models.

DOI: 10.4324/9781003268697-4

## The Mainstream Conception of Sustainability Indicators and its Philosophical Assumptions

In standard scientific parlance, an indicator is a piece of information about the static or dynamic properties of entities. Indicators "summarize or otherwise simplify relevant information, make visible or perceptible phenomena of interest, and quantify, measure, and communicate relevant information" (Gallopin, 1996, p. 108). *Sustainability* indicators, then, provide summary information about our progress toward sustainability. Because the socio-ecological systems are complex, sustainability indicators and indices assess progress using a limited number of measurable parameters. They communicate partial information about the "state, dynamics and drivers affecting human-environmental systems" (Wu & Wu, 2012). As a way of "operationalizing the concept of evidence-based policy" (Sébastien et al., 2014, p. 318), sustainability indicators are supposed to give members of the public, politicians, and governmental and non-governmental agency personnel the ability to answer the question whether the situation is getting better or worse.

An indicator is not a 'fact' like other types of information such as data and statistics. Like 'pass' and 'fail' when attached to a test score, sustainability indicators are *evaluative* labels. Inherently, they involve a three-place relation between data, claim and background assumptions about the desirability of situations. However, viewing indicators as "objective empirical measures" (Jacques, 2021, p. 96) that can directly inform policy ineluctably pulls them back into two-place relations between evidence and the evaluative assessment. This links them strongly to the "direct empiricist" conception of data and evidence in the philosophy of science. Direct empiricism conceptualizes data as neutral facts. Direct perception is treated as the exemplar. Observation of perceptual data – colors, shapes, distances, etc. – is taken to unproblematically and directly serve as evidence for the claim that, for instance, I am seeing trees in my backyard.[3] On this view, data are treated "as windows on the world, as reflections of reality, without any art, theory, or construction interfering with that reflection. . . . [They] represent the naked or unmediated truth about [a] particular aspect of the world" (Lloyd, 2012, p. 392). Data are just "elements of reality," and any meaning added during processing is "tampering" (Bokulich, 2020, p. 793). At the ontological level, direct empiricism "myopically flatten[s]" the objects of the claims "into the surface of evidence" (Feigl, 1958, p. 98). Observational data become synecdoches for things.[4]

In the context of policy, direct empiricist conceptions of data are associated with a rationalistic, linear, and instrumental conception of the role of science-based knowledge. The underlying mindset is that better policy processes will "automatically follow as robust, data-driven, objective, and value-free evidence is made available for policy makers" (Sébastien et al., 2014, p. 318). Data do nothing more than reveal options and help us make informed decisions. Here, objective data informs policy directly, leading to value-free, objective policy making. Data are windows on the policy world. The overall effect is to place evidence outside social, cultural, and political influences (Latour, 1993). Data and evidence – and the scientific, technological, and natural processes they are linked to – become privileged forms of knowledge insulated from power.

Direct empiricism makes sense of both the 'objective data' and 'policy relevant' aspects of mainstream sustainability discourse about indicators. Indicators are often seen as "data carriers" (Merino-Saum et al., 2020), "simple listings under themes" (Bell & Morse, 2018), and "a set of boxes to be ticked" (Sari et al., 2018, p. 423). They are a "type of universal language" (Murray et al., 2022) connecting all aspects of sustainable behavior. They have meaning "independently from the context, purpose and logics behind their use" (Merino-Saum et al., 2020). This is solidified in data-intensive approaches, where "the data are understood to have a meaning that can be uncovered without the need for human interpretation" (Asokan et al., 2019, p. 958).

A good example here is the interpretation and use of the United Nations' Sustainable Development Goals (SDGs). The list of SDGs assumes that the meanings of 'quality education,' 'responsible consumption and production,' 'life on land,' etc. exist prior to any inquiry. Researchers and policy makers are directed to find data on those features of society and nature. The UN recognizes that actual meaning of each term will change from place to place, but the overall meaning of sustainability has been charted prior to any investigation.

Philosophers interested in sustainability often adopt similar readings of indicators as unproblematic windows on the world. Curren and Metzger (2017, p. 230) argue that the planetary boundaries approach "is a measure of ecological sustainability" in ways that indicator-based approaches such as the Ecological Footprint Analysis (EFA) are not. For Norton (2015), indicators are direct or indirect, actual, or imagined "empirical measures of progress in providing or protecting a social value" (p. 120). Indicators directly reflect and project realities chosen by communities.

As objective empirical measures, sustainability indicators are often associated with three other theses loosely associated with direct empiricism. The first is that data is independent of, or separable from, other data. Some international organizations recommend that each UN Sustainable Development Goal be audited separately (Sari et al., 2018). Urban sustainability components are commonly viewed as "detachable pieces" (Merino-Saum et al., 2020) in sustainability analyses. Overall, the position seems to be that hard facts have and need no relation to other facts for them to be meaningful.

The second is substitutability, or replaceability. Indicators collect different features of a situation into a standard measure to report on the overall state of the system. This encourages the idea that features are interchangeable. For example, the Human Development Index (HDI) standardizes indices of life expectancy at birth (a proxy indicator for health care and living conditions), mean years of schooling for adults aged 25 years and more, and gross national income per capita (UNDP, n.d.). "Once standardized, the three partial indices are added and divided by 3 to provide an average. Summing the components in this way implies, [*sic*] replaceability, that is increases in education can compensate for declines in life expectancy" (Morse, 2004, p. 89). Although tweaked regularly, this methodology has been employed since the inception of the HDI. Many indices are multi-dimensional and do not employ substitutability directly. However, the pull of providing a simple assessment about the state of the system often leads to ignoring trade-offs among the features as separate, incomparable measures of well-being (Hirsch et al., 2010).

The third is reductionism. Because they "attempt to encapsulate complex and diverse processes in a relatively few simple measures," sustainability indicators are "a classic reductionist set of tools based on quantification" (Bell & Morse, 2008, p. 42). Many earlier methodologies used to assess sustainability were extremely reductionistic. They focused on single dimensions such as economics, single time horizons (often set by political constraints), and single measurable indicators. Development indicators that focused on economic measures such as GDP per capita are good examples. While multi-criteria approaches are now more widely employed, some rigorously reductionistic approaches remain. Biophysical approaches to sustainability are good examples. Farley and Smith (2020) employ the thermodynamic framework of Georgescu-Roegen (1971) and others to argue that the limits of the environment are set by the production of entropy. Indicators such as decreasing glaciers and snow cover, retreating sea ice, warming, more acidic seas, etc. "support the message of the limits argument" (Farley & Smith, 2020, p. 23). Better and more rigorous policies will ensue if we pay attention to indicators that illuminate the thermodynamic rules that govern natural systems. The planetary boundaries framework (Steffen et al., 2015) also approaches indicators in a reductionistic biophysical manner. Only the boundaries associated with ozone depletion, loss of biosphere integrity, etc. are needed to assess whether humanity can continue to develop and thrive. While robust sets of indicators derived from planetary boundaries are as yet only envisioned, Li et al. (2021) propose formulating "absolute environmental sustainability indicators" that respect biophysical limits and improve human well-being (where such improvement is gauged by whether activity remains within the biophysical limits).

As in the evidence-based policy movement more generally (Clarence, 2002; cf. Persson et al., 2018), indicators as neutral and objective tools are interpreted as allowing one to "inject hard data . . . into the policy-making process" (Holden, 2001, p. 217). They are "often viewed as conceptually simple tools" (Kourantidou et al., 2020). In many contexts, their selection and development are "largely perceived as technical, bureaucratic, and scientific" (Muhl et al., 2022). These "technical policy tool[s]" permit "monitoring and the subsequent 'steering' of policy" (Rydin et al., 2003, p. 581; cf. Merino-Saum et al., 2020; Hezri, 2004, Sébastien et al., 2014; Elgert, 2018). This "informational managerialist" (Mol, 2006, p. 506) attitude supports a "traditional rational-positivist perspective" of the policy process, in which "indicators are expected to 'close down', enabling better management and control by providing robust, accurate, quantitative and unambiguous information for the purposes of political advocacy and day-to-day policymaking" (Lehtonen et al., 2016).

Examples here are numerous. Even though indicators developed from planetary boundaries have yet to be constructed, the framework is self-consciously proposed as a context for policy making (Stockholm Resilience Centre, n.d.). The Driver-Pressure-State-Impact-Response (DPSIR) framework is used widely in policy work (Stanners et al., 2007). The framework structures and organizes measurements of, say, increased fertilizer use in agriculture and increased nitrate concentrations in surface waters into categories of drivers, pressures, etc. The framework of associations helps policy makers "understand the meaning of information in indicator reports" (Tscherning et al., 2012). DPSIR is seen as providing robust, unambiguous information to support decision-making.

The managerialist mindset often remains even when more people are brought into the discussion. Instead of academic or agency experts steering the process alone, today they often serve as one of many different stakeholders in participatory sustainability assessment frameworks. Several such frameworks are available. Stakeholder involvement occurs in varying degrees and at various stages of the sustainability assessment. Different methodologies are used to allow participants to designate what they consider to be the most appropriate solution (Gasparatos et al., 2008). Despite the differences in stakeholder involvement, the indicators usually play the same roles as before, "providing feedback on the effects of policies, identifying and describing unsustainable evolutions, co-constructing visions and evaluating pathways towards desired societal change" (Lehtonen et al., 2016).

I believe such a view is found in Bryan Norton's approach to sustainability. As he conceives it, the project for members of the community is to "sort through and choose sustainability indicators appropriate to them and their values" (Norton, 2005, p. 369). "[E]ach community must express its sense of place in the process of choosing indicators that highlight processes important to the [sic] its traditional and evolving values" (Norton 2015, p. 129). In the deliberative process in which the values are expressed, the sciences are handmaidens, "offering ways to formulate indicators and criteria that are both precise and expressive of social values" (Norton, 2015, p. 365). Communities should identify the key indicators that support their values, not argue about the language used to express the values or the grounding used to support them. This process operationalizes the public interest for a given community.

The process of deliberation will be "hard work" (Norton, 2015, p. 129). But note that indicators are directly reflective of values. In this sense, they serve as 'windows on the world' of social values. And the indicators serve as direct inputs into the deliberative process, serving as direct links to values. The indicators serve as neutral (in the sense of 'acceptable to all') language for the construction of the vision for change. All of this appears to reinforce the assumption that indicators provide unproblematic feedback into the policy process.

## A Complex Empiricist Approach

Despite the ubiquity of indicators, their influence on decision-making is "largely insufficient" (Waas et al., 2014, p. 5519; cf. Hák et al., 2018; Holden, 2013; Mascarenhas et al., 2015; Turnhout et al., 2007; Verma & Raghubanshi, 2018). The haphazard use happens for many reasons for including the quality of the indicators, the people using them and the policy contexts in which they are produced and used (Sébastien et al., 2014, p. 322). But I think it is plausible to argue that direct empiricist framings of indicators as data contribute to the problem. Seeing indicators as pieces of data strips them of their contexts. As Wellstead et al. (2016) argue, this embodies a structural-functionalist view where problems in the socio-ecological realm are assumed to generate fitting policy responses. This ignores several policy-relevant factors such as the division of powers among agencies, the impact of policy legacies, and blame-averse behaviors. Further, it leads to generic management prescriptions such as ecosystem management.

A thick approach that locates indicators in their theoretical, methodological, socio-political, and normative contexts offers a richer conception. In complex empiricism (cf. Suppes, 1962; van Fraassen, 2008; Giere, 2006; Lloyd, 2012), all evidence is always already mediated by non-trivial, theoretically informed assumptions. Another way of saying this is that evidence is always a three-place relation between a fact, a hypothesis and the reasons that show why the evidence is relevant to the hypothesis (Cartwright, 2013; Longino, 1990). Van Fraassen's (2008) example of weather reports offers a good example of the approach. Weather data seems simple and direct. However, even something like a graph of yesterday's temperature over the course of the day is a representation. Temperatures are collected from many different stations at discrete moments during the day. They are then statistically processed to arrive at a "smoothed-out summary" (van Fraassen, 2008, p. 166) that is the graph. The smoothing is needed because the temperatures recorded at the multiple stations are subject to multiple cloud, tree, and land cover effects. For example, I have a stand of trees all around my house, so the temperatures on my outdoor thermometer in the summer are nearly always a few degrees cooler than the nearby stations on my weather app. The outdoor sensor is accurate, at least in the sense that it agrees with the temperature inside my house when I bring it in. Once one sees the smoothing process, one sees that, rather than an unmediated picture of reality, what one has is a claim that "the object looks like this in this measurement set-up" (van Fraassen, 2008, p. 167). Assumptions about the relations among the individual data points, about the validity of the smoothing, etc. are essential to producing the 'fact' of the graph. Winsberg (2018) offers an extended argument that complex empiricism makes sense of climate discourse.

Lest one think that the individual data point is the 'real' datum, note that it is tied to a particular measurement set-up, which is, in turn, tied to our "informational and work requirements" (Star, 2010). It is relevant because I want to know whether to wear a hat or because I need detailed, fine-grained information for a study about temperatures at particular places and particular times. Individual data points are still *only* data about an object in that measurement set-up. It looks *that way* given our interests in the information. Data is always theory-laden, embedded in interests and assumptions.

Data fusion and framing are two of the many ways theory-ladenness arises in sustainability discourse. Data fusion, also known as data integration, refers "to a broad family of methods for combining heterogeneous data sources into a coherent and improved data product" (Bokulich, 2020, p. 799).[5] Data is collected in various units, over various times, with respect to various reference points, etc. Moreover, a given feature may have been measured in different units at different points in time and place. Finally, records of the same feature at different locations and times can be incomplete and fallible. These heterogenous elements must be processed into a common measure.

As an example, take the United Nations Environment Programme's initial Global Environment Outlook (UNEP, 2007). To generate a conclusion about the overall state of the environment, the report combined data on the following features: regional changes in population, gross domestic product, primary energy consumption, energy intensity, agricultural production (in the form of maize production), caloric intake, total water withdrawal, changes in land use and cover, and habitat loss (UNEP, 1997; cf. Dahl, 2007, p. 166). Note quickly that much of the data collected for a particular feature are themselves models of other data. For instance, note that agricultural production is referenced to maize production only. But more importantly for my purposes, none of the individual data points provide a direct claim about the overall state of the environment or about whether it is improving. The different features are measured with different units, over different time and spatial scales and with different degrees of reliability, to name just a few of the varying features of the measurements. The data must be integrated into a single claim about the state of the environment. Similar remarks apply to the creation of indicators in other domains of sustainability. Holden (2001) and Merino-Saum et al. (2020) provide good discussions in the field of urban sustainability.

Second, indicators involve framing. Framing involves far more than the routinely recognized point that indicators are subjective because they involve judgments about what phenomena need to be measured and about which features of the phenomena should be included in the reductionistic indicator (cf. Meadows, 1998; Stanners et al., 2007; Bell & Morse, 2018;

Molden & Dahl, 2007). This structures the question as one about different windows onto the *same* phenomena. However, framing affects not only the windows but what one sees as the phenomena and the problem. Among other things, frames define boundaries, entities, scales, and a focus on processes inside the boundaries versus those that cross them (Kay, 2008). Frames determine "what counts as a fact and what arguments are taken to be relevant and compelling" (Wesselink & Warner, 2010; cf. Schoen & Rein, 1993). A classic example is research on agricultural production and food security challenges in the 1990s caused by road construction in the 1940s in the Ucayali region of Peru. Research agencies for years framed research in terms of cattle management, slash-and-burn agriculture, and deforestation. In the late 1990s, a new team of researchers generated a different systems description. They found that fish was the primary source of protein and that cattle did not play an important role in food production or income generation. Rather than slash-and-burn agricultural practices, annual crops provided most of the food and income. And the illegal coca crop was cultivated over an area larger than the two main legal crops of plantains and bananas combined. Rather than 'deforestation,' the research team focused on ways timber extraction employs large numbers of people, generates the greatest amount of revenue and is a main export product (Murray et al., 2008). The two descriptions refer to different facts, different entities, and different drivers, and they direct our attention to different processes occurring inside and outside the boundaries. Lest one think that the newer description is the 'correct' one, one should remind one's self that any description is a limited mental representation and thus that "any situation should be described by a number of systems descriptions" (Kay, 2008, pp. 16–17).

Once processes of data fusion and framing (and other kinds of theory-ladenness not explored here) are acknowledged, indicators begin to look very different than empirical measures that steer policy. As evidence, indicators become "arguments, ideas and expectations that particular actors mobilize regarding sustainability issues" (Merino-Saum et al., 2020). Rather than measuring an objective, fully defined concept of sustainability, developing and using indicators "constitutes a process through which the concept . . . acquires content and is defined" (Merino-Saum et al., 2020). In turn, this contextual slant foregrounds the processes of fusion and framing. Theory-laden indicators force one to recognize that any index, no matter how comprehensive, is tied to judgments about how a given problem is framed, e.g., as urban or agricultural or ecological sustainability. Framing means selecting features as important. And since some fusion is inevitable as the data is aggregated into a report about the system, thinking of indicators as theory-laden calls attention to the processes of fusion and the judgments involved in them.

This thick notion of sustainability data works against separability, substitutability, and reductionism. Seeing indicators as models situates "an individual indicator within a broader network of information" (Sébastien et al., 2014, p. 318). One looks at "how an individual indicator . . . contributes to the 'entire story' as well as how it articulates with the remaining indicators within the same set" (Merino-Saum et al., 2020). Context and system come to the fore. Likewise, rather than assuming disparate phenomena can be studied with a common measure, which erases context and values (O'Neill et al., 2008, p. 194), the thick notion directs attention to questions about how and why the phenomena are important to us. Rather than assuming more education can be traded off equally for a shorter life expectancy, we must ask what they mean to us. Theory-ladenness also issues a caution against reductionism. For instance, rather than using only reduction in educational disparities to measure gender equality and women's empowerment, as in the Millennium Development Goals, precursors to the current SDGs, one includes several other important dimensions of equality and empowerment such as sexual, reproductive, economic, political, and legal rights, all of which affect access to education and thus whether educational disparities can be reduced (Sen & Mukherjee, 2014). As another example, biophysical frameworks such as planetary boundaries simply posit in advance that the biophysical limit is the determinative factor of a sustainability claim. However, there is no built-in warrant that the biophysical limit is the most important feature such that other dimensions of sustainability should be read through it.

By calling attention to the context within which indicators get their meaning, an "indicators as models" approach also emphasizes the presence of trade-offs, path dependencies and policy resistance. Framing calls particular attention to trade-offs in the construction and use of indicators. The construction of every indicator involves "decisions to exclude and include" (Bell & Morse, 2018). For example, when considering the drivers of urban (un)sustainability, US and Canadian indices focus more attention on social issues while Chinese indices focus more on matters of technology and the economy. Other trade-offs involve choices to represent systems with a few, simple indicators or more indicators that represent various sub-systems and different dimensions of sustainability. Yet others necessitate decisions about specificity versus general comparability and detail and credibility versus understandability for users (Merino-Saum et al., 2020).

Trade-offs are also numerous when indicators are used. Progress along one dimension of sustainability can conflict with progress along others. In the case of trying to satisfy the UN's SDGs independently, "as countries manage to lift millions out of poverty and provide much-needed health care, the demands on affordable and clean energy currently rises at a rate that jeopardizes progress regarding the Agenda 2030" (Kroll et al., 2019; cf. Fuso Nerini et al., 2019). Complex empiricism expects trade-offs among the features measured by indicators. "Adopting a systemic perspective, which emphasizes the performative, indirect and largely uncontrollable impacts of indicators, draws attention to the numerous unavoidable trade-offs between various indicator roles and functions – trade-offs that stem from the nature of indicators as boundary objects" (Lehtonen et al., 2016). All these trade-offs in construction and use are foregrounded when the theory-laden, contextual nature of indicators is emphasized.

Likewise, an 'indicators as model' approach highlights the existence of path dependencies, or lock-in (Hukkinen, 2004; Pradhan et al., 2017; Pupphachai & Zuidema, 2017). Path dependencies are features of systems, and indicators taken as separate data points do not show which properties of the system are favored due to existing technological, institutional, and/or behavioral infrastructures. This can lead to further entrenching the existing system. A focus on indicators for smart cities often entrenches existing unsustainable behaviors (Buzási & Csizovszky, 2022). A push toward more technologically efficient vehicles can entrench the culture of the automobile (Geels, 2012). Improving the accessibility and quality of energy services, which is UN SDG 7, accepts the demand for energy and, given existing technological, institutional, and behavioral infrastructure, can lead to locking in fossil fuel use (Seto et al., 2016).

There is also some reason to believe that an 'indicators as models' approach could more fully capture the phenomena of policy resistance, the phenomenon whereby policy interventions are defeated by the system's response to the intervention itself. (I say 'could' because the effects of many interventions are not yet fully known.) For example, many cities and governments use indicators developed separately in the sectors of energy provision and mobility. Decarbonization efforts focused on reducing greenhouse gas emissions often stymie other sustainability goals. Home-based solar panels used to charge electric cars produce local, cheap energy but perpetuate problems of car-based urban mobility and discourage walking and use of public transportation. Such efforts may also result in energy and transport poverty among marginalized populations (Payakkamas et al., 2023). Similar resistances have been reported in moving Sweden to a fully bio-based economy (Bennich et al., 2018). By locating any investigation theoretically and contextually, a complex empiricist approach directs attention at assumptions about separateness and interactions among sectors.

Theory-ladenness also affects views about the use of indicators in the policy process. Rather than neutral tools, they become tools with social, political, and moral dimensions. As Winner (1980) would put it, they become artifacts that have politics. To see this, it is worthwhile to lay out some of the negative consequences of viewing indicators as neutral, technical tools that enable better management and control. For one, it often distorts what counts as knowledge. Practical experience, in the form of local knowledge about complex and site-specific relationships developed over time, can easily become interpreted as anecdotal, weak, or biased. This often devalues Indigenous ways of knowing (Persson et al., 2018; cf. Kourantidou et al., 2020; Muhl et al., 2022).

It also distorts policy making. Consider countries' recent reports of greenhouse gas emissions to the United Nations Framework Convention on Climate Change (UNFCCC). Many countries offset fossil fuel emissions by claiming that carbon is absorbed by land. UN rules allow countries to report this as a single, hard fact. The issue here is that there is no single answer to the question of how much carbon dioxide is absorbed by land. It is routinely assumed that mineral-rich grassland soils are neutral in their uptake of carbon, despite evidence to the contrary. The give and take of carbon and nitrogen between land and the atmosphere is incompletely known. Different numbers are generated if one counts only deliberate action such as the planting of new forests or also allows natural processes of absorption into the calculation. In sum, "[a]ny number produced is filtered through a large number of assumptions about processes that are incompletely known" (Mooney et al., 2021). More generally, seeing data as hard facts that steer policy directly underestimates the impact of politics, judgment, and impromptu strategy-building in science, in policy making and in the interaction between the two (Jasanoff, 2004; Sébastien et al., 2014; Shrader-Frechette, 2013; Wesselink et al., 2013).

In addition, the thin conception erases both values and politics. It tends "to reduce value conflicts and normative debates to presumably neutral and commonly agreed numbers perceived as incontestable facts" (Lehtonen et al., 2016). This depoliticization erases questions of power. A rich example is the 'fact' of minimum flow requirements (MFR) on the Garonne river of southwestern France. In 1996, a watershed agency institutionalized MFR to objectives that would reduce water deficits on the river and allow action plans to be formulated. As Fernandez (2014) argues however, actors use the fact of MFR to "erase history and naturalize rationales" (p. 258) regarding access to the water. Historically, the first problematization of the river level occurred during fierce contestation over development in what was, at the time, considered an economically backward area. In the 19th century, civil engineers in government and industry set minimum flow requirements in attempts to stop some projects of canal construction and river improvement and favor others. The MFR adopted then remains, for the most part, to this day. Today, actors from nuclear, hydroelectric, agricultural, and fishing industries as well as local governments (which use the river for producing drinking water) appeal to it as a hard fact to justify various projects. The "a-historical and a-political dimensions that the MFR pretends to have tend . . . to increase the system path-dependency" (Fernandez, 2014, p. 269). That is, political actors use the fact to entrench their positions rather than addressing larger questions about why river flow is an issue.

This is not an isolated example. A conception of planetary boundaries as rigid "renders debate about values irrelevant by asserting that violating [the boundaries] would cause intolerable harm to human prospects" (Randall, 2021). Hansson et al. (2019) trace the depoliticization in the use of the UN SDG indicators, and Merino-Saum et al. (2020) argue that urban sustainability indicators using the SDGs "largely ignore the disparities component, thereby disregarding political ecology concerns about the access to and the management of natural resources." Miller (2019) notes that 'smart cities' discourse

deploys information and communications technologies to create standards and indicators of urban sustainability that use "the cloak of urban science and big data to appear apolitical as they attempt to advance sustainability goals."
In contrast, the 'indicators as models' approach readily captures the various normative dimensions of indicator construction and use. For one, it bakes descriptive and semantic normativity into indicators. By their very nature, indicators contain "a more or less precise reference point (e.g., a target, a benchmark, a threshold, a range or simply an orientation) *through which the data might be properly considered*" (Merino-Saum et al., 2020, emphasis added). As a certain type of model, indicators come to have meaning only through an external reference, one that is taken to show that things are getting better, worse or staying the same. It also recovers methodological normativity. Rather than rigid markers, they are pieces of information that involve judgment. For instance, in the case of planetary boundaries, one sees their fuzziness and, as a result, the uncertainty of judgments about 'safe' and 'unsafe' zones and about the processes of collapse and recovery (Randall 2021; cf. Ramsey 2017). Finally, it pulls in moral normativity directly. We judge that a situation is good or bad for us, the environment, or some socio-ecological system.

All this opens up normative reflection on the tools themselves. As tools, indicators actively shape understandings and actions in intended and unintended ways. As Winner (1980) argued some time ago, artifacts – i.e., tools – have politics. Focusing on what indicators *do* as tools directs attention to how they alter relationships among natural, social, and socio-ecological systems.

Thinking of indicators as 'boundary objects' highlights this performative aspect. A boundary object is a set of material and processual work arrangements. The necessity of cooperation and collective problem solving among heterogeneous actors and their multiple, often conflicting viewpoints requires some sort of process for resolving the conflict. This is accomplished sometimes by carving out domains of authority (Gieryn, 1983) and sometimes by cooperating across the divide (Star, 2010; Star & Griesemer, 1989). These are complementary processes, often occurring intentionally or not at the same time. Boundary objects involve interpretive flexibility, are tied to work requirements, and dynamic negotiation about their meaning.

As theory-laden measurement set-ups intended to be used by multiple groups for multiple reasons, indicators are quintessential boundary objects. Indicators are boundary objects because they provide interpretive flexibility among disparate groups (Brand & Jax, 2007; Garmendia et al., 2016; Oberlack et al., 2019; Termorshuizen & Opdam, 2009). However, this is a minimal sense since 'better communication' barely – if at all – invokes processes of negotiation over contested meanings and uses. It returns indicators to the status of neutral tools (of communication), forgetting the ways groups can talk about the 'same' object but interpret its meaning in very different ways. The indicators of gender equality and women's empowerment in the Millennium

Development Goals can mean reduction in educational disparities to some, while to others it involves sexual, reproductive, economic, political, and legal rights.

Two other components of boundary objects tie them even more strongly to their performative and thus multiply normative nature. First, boundary objects are non-arbitrary structures arising from information and work requirements. For example, as a repository, a library allows things to be removed without changing the structure of the whole and is particularly well suited to allowing private investigations (Star & Griesemer, 1989). In forest management, academics from climate science, hydrology, forest ecology, fire ecology and social psychology as well as managers from federal and state agencies and NGOs interpret the significance of indicators tied to global circulation models, regional climate and water models and regional vegetation and fire models in different ways based on what they see as necessary to produce their results (Blades et al., 2016). As information and measurement devices tied to specific conceptualizations and interpretations of situations, indicators are structured by practice.

In addition, boundary objects involve dynamic negotiation. They allow cooperation, but users maintain the vaguer identity while making it more specific, more tailored to local use in their 'home' group. If the groups are cooperating without consensus, they tack back and forth between vague and more tailored uses of the objects. In this negotiation, actors discipline the objects. "[P]eople (often administrators or regulatory agencies) try to control the tacking back-and-forth, and especially, to standardize and make equivalent the ill-structured and well-structured aspects of the particular boundary object" (Star, 2010, pp. 613–614). Fernandez's (2014) discussion of the minimum flow requirements on the Garonne provides a good example. Some minimum flows optimized the dilution of ammonia, which channels finances toward dams and wastewater treatment facilities. Others focused on irrigation based on 'natural' flows, which leads to a focus on agricultural practices independently of fertilizer use. Yet others focused on the effect of dams on the river or on drinking water quality.

As Turnhout (2007), Lehtonen et al. (2016) and Sébastien et al. (2014) argue, seeing indicators as epistemic devices involving flexibility, practice-based structures and dynamic negotiation implies that indicators as tools shape and are shaped by use and by work and information requirements and by the social and political negotiations required to allow cooperation without consensus. Derickson (2009), Kourantidou et al. (2020), and Muhl et al. (2022) provide analyses supporting this point.

Overall, a performative reading of indicators as boundary objects emphasizes the idea that indicators are *in* society, culture, and politics. By analogy to the argument about the meaning of sustainability developed in Chapter 2, there can be no single meaning of a sustainability indicator. Different contexts provide different meanings. And one of those contexts is policy. Rather than a linear uptake of knowledge by policy makers, indicators become one piece of information in a process of constant negotiation and adjustment.

Just as the height of the bridges kept buses off the Long Island Expressway, indicators have the power to affect our behavior and shape our view of society. "In our performance-oriented world, measurement issues have taken on increased importance: what we measure affects what we do" (Stiglitz, 2009, p. 1). Once they are affecting our behavior, they tend to continue to do so. In the Human Development Index, changes tend to occur around the edges rather than asking new questions with new approaches (Morse, 2004). And they do more than just alter our personal behavior. They can be used to control them. As tools in use by powerful actors, indicators can be seen as "as normative systems for control, intervention, inclusion and exclusion. ... [They] shape legitimate ways of being and acting" (Hansson et al., 2019, p. 219; cf. Scott, 1999). Even if indicators are developed from the ground up by local groups and include multiple dimensions of sustainability, "the question remains whether indicator measurement of any description may lead to more than a quest for constancy, order, and administrative efficiency and the concurrent delegitimization of the variable, specific, nonstandardized logic of ecosystems and human communities" (Holden, 2001, p. 232). Searching for systematic mechanisms that link indicator use in policy "may hide rather than illuminate the context dependent dynamics at play" (Sébastien et al., 2014, p. 235). Indicators can be used to hamper dialog and deliberation, and legitimize or reinforce prevailing power structures (Lehtonen et al., 2016). In contrast, a contextual, 'indicators as models' approach "weaves power and knowledge together" (Muhl et al., 2022). By thinking about how tools affect us and our ability to act, we address "the politics inherent in the processes of indicator development, use, and influence" (Sébastien et al., 2014, p. 235). As performative tools, indicators are no longer neutral.

## Notes

1   On JSTOR, Philosopher's Index and the History of Science, Technology and Society databases, January 2023 searches on "sustainability indicators" and "evidence" or "data" returned fewer than 30 items each. Most articles assessed how corporations use indicators in their sustainability reports. A few assessed the use of indicators as markers of well-being in policy contexts. Miller (2005) argues that using sustainability indicators as replacements for conventional welfare measures transforms relations between knowledge and the state. This is relevant but largely addresses the effects of indicator use rather than their nature or the claim that indicators can serve as direct inputs to policy.
   The closest body of literature comes from STS perspectives on governance and governmentality. I use these in the chapter. The literature on quantification (cf. Porter, 1996) addresses the societal effects of indicators more broadly, but I am not aware of much discussion that connects this issue specifically with questions about sustainability indicators as evidence.
2   From a conventional perspective that holds data has no meaning on its own and becomes evidence only in the presence of an argument, everything I say regards

evidence. Indicators are evidence *of* a state or process. However, indicators are treated as data, particularly in policy contexts. Even more, as I note later complex empiricism holds that 'raw' data are linked to arguments because they are inevitably laden with assumptions and theory. For these reasons, I often use the two terms interchangeably.

3 As philosophers and philosophers of science know, even simple examples like this are not great examples of direct empiricism. At the very least, the visual data is processed into a claim about objects using assumptions about 'normal' perceivers and 'normal' perceptual conditions. And if cognitive scientists like Marr (1982) and Gibson (1979) are correct, quite a bit of inferential processing in the brain is needed to convert the raw percepts into an object, even during 'normal' perception.

4 This was an important piece of the logical empiricist program in the philosophy of science. Some later positivists tried to break away from the strictures on evidence, but the pull of epistemic convention was very strong (Godfrey-Smith, 2021). Versions of the approach to remain in common parlance among empiricist-minded philosophers. For example, it is not uncommon to read and hear that 'trilobites are fossils' is a simple fact. This is asserted despite the heavily theory-laden nature of the claim: a pattern in a rock must be seen as a no-longer living animal, that individual has to be placed into a taxonomy, the pattern has to be associated with the taxonomy, etc.

5 Bokulich (2020) provides what she regards as a preliminary taxonomy of the processes by which data become theory-laden, elucidating how understanding the processes helps us develop a more adequate philosophy of data. In addition to fusion, Bokulich mentions conversion, correction, interpolation, scaling, assimilation, and synthetic data produced from simulation models. I believe many are used – often at the same time – in the construction of indicators and indices. I leave this as a future project. Morse (2004) discusses the various processes used in the creation of indicators.

## References

Asokan, V. A., Yarime, M., & Onuki, M. (2019). A review of data-intensive approaches for sustainability: Methodology, epistemology, normativity, and ontology. *Sustainability Science, 15*(3), 955–974. https://doi.org/10.1007/s11625-019-00759-9

Bell, S., & Morse, S. (2008). *Sustainability indicators: Measuring the immeasurable?* 2nd ed. Earthscan.

Bell, S., & Morse, S. (2018a). Sustainability indicators past and present: What next? *Sustainability, 10*(5), 1688. https://doi.org/10.3390/su10051688

Bell, S., & Morse, S. (Eds.) (2018b). *Routledge handbook of sustainability indicators.* Routledge.

Bennich, T., Belyazid, S., Kopainsky, B., & Diemer, A. (2018). Understanding the transition to a bio-based economy: Exploring dynamics linked to the agricultural sector in Sweden. *Sustainability, 10*(5), 1504. https://doi.org/10.3390/su10051504

Blades, J. J., Klos, P. Z., Kemp, K. B., Hall, T. E., Force, J. E., Morgan, P., & Tinkham, W. T. (2016). Forest managers' response to climate change science: Evaluating the constructs of boundary objects and organizations. *Forest Ecology and Management, 360*, 376–387. https://doi.org/10.1016/j.foreco.2015.07.020

Bokulich, A. (2020). Towards a taxonomy of the model-ladenness of data. *Philosophy of Science, 87*(5), 793–806. https://doi.org/10.1086/710516

Brand, F. S., & Jax, K. (2007). Focusing the meaning(s) of resilience: Resilience as a descriptive concept and a boundary object. *Ecology and Society, 12*(1), 23. https://doi.org/10.5751/es-02029-120123

Buzási, A., & Csizovszky, A. (2022). Urban sustainability and resilience: What the literature tells us about "lock-ins"? *Ambio, 52*(3), 616–630. https://doi.org/10.1007/s13280-022-01817-w

Cartwright, N. (2013). *Evidence: For policy and wheresoever rigor is a must.* London School of Economics.

Clarence, E. (2002). Technocracy reinvented: The new evidence based policy movement. *Public Policy and Administration, 17*(3), 1–11. https://doi.org/10.1177/095207670201700301

Curren, R., & Metzger, E. (2017). Preserving opportunity: A précis of living well now and in the future: Why sustainability matters. *Ethics, Policy & Environment, 20*(3), 227–239. https://doi.org/10.1080/21550085.2017.1374000

Dahl, A. L. (2007). Integrated assessment and indicators. In T. Hák, B. Moldan & A. L. Dahl (Eds.), *Sustainability indicators* (pp. 163–176). Island Press.

Derickson, K. D. (2009). Gendered, material, and partial knowledges: A feminist critique of neighborhood-level indicator systems. *Environment and Planning A: Economy and Space, 41*(4), 896–910. https://doi.org/10.1068/a40255

Elgert, L. (2018). Rating the sustainable city: 'measurementality', transparency, and unexpected outcomes at the knowledge-policy interface. *Environmental Science & Policy, 79,* 16–24. https://doi.org/10.1016/j.envsci.2017.10.006

Farley, H., & Smith, Z. (2020). *Sustainability: If it's everything, is it nothing,* 2nd ed. Routledge.

Feigl, H. (1958). The 'mental' and the 'physical.' In H. Feigl, M. Scriven & G. Maxwell (Eds.), *Concepts, theories and the mind-body problem* (pp. 370–497). Minnesota Studies in the Philosophy of Science 2. University of Minnesota Press.

Fernandez, S. (2014). Much ado about minimum flows... Unpacking indicators to reveal water politics. *Geoforum, 57,* 258–271. https://doi.org/10.1016/j.geoforum.2013.04.017

Fuso Nerini, F., Sovacool, B., Hughes, N., Cozzi, L., Cosgrave, E., Howells, M., Tavoni, M., Tomei, J., Zerriffi, H., & Milligan, B. (2019). Connecting climate action with other sustainable development goals. *Nature Sustainability, 2*(8), 674–680. https://doi.org/10.1038/s41893-019-0334-y

Gallopín, G. C. (1996). Environmental and sustainability indicators and the concept of situational indicators. A systems approach. *Environmental Modeling & Assessment, 1*(3), 101–117. https://doi.org/10.1007/bf01874899

Gallopin, G. C. (2018). The socio-ecological system (SES) approach to sustainable development Indicators. In S. Bell & S. Morse (Eds.), *Routledge handbook of sustainability indicators* (pp. 329–346). Routledge.

Garmendia, E., Apostolopoulou, E., Adams, W. M., & Bormpoudakis, D. (2016). Biodiversity and green infrastructure in Europe: Boundary object or ecological trap? *Land Use Policy, 56,* 315–319. https://doi.org/10.1016/j.landusepol.2016.04.003

Gasparatos, A., El-Haram, M., & Horner, M. (2008). A critical review of reductionist approaches for assessing the progress towards sustainability. *Environmental Impact Assessment Review, 28*(4–5), 286–311. https://doi.org/10.1016/j.eiar.2007.09.002

Geels, F. W. (2012). A socio-technical analysis of low-carbon transitions: Introducing the multi-level perspective into transport studies. *Journal of Transport Geography, 24,* 471–482. https://doi.org/10.1016/j.jtrangeo.2012.01.021

Georgescu-Roegen, N. (1971). *The entropy law and the economic process.* Harvard University Press.

Gibson, J. J. (1979). *The ecological approach to visual perception.* Houghton-Mifflin.

Giere, R. (2006). *Scientific perspectivism.* University of Chicago Press.

Gieryn, T. F. (1983). Boundary-work and the demarcation of science from non-science: Strains and interests in professional ideologies of scientists. *American Sociological Review, 48*(6), 781. https://doi.org/10.2307/2095325

Godfrey-Smith, P. (2021). *Theory and reality,* 2nd ed. University of Chicago Press.

Hansson, S., Arfvidsson, H., & Simon, D. (2019). Governance for sustainable urban development: The double function of SDG indicators. *Area Development and Policy, 4*(3), 217–235. https://doi.org/10.1080/23792949.2019.1585192

Hezri, A. A. (2004). Sustainability indicator system and policy processes in Malaysia: A framework for utilisation and learning. *Journal of Environmental Management, 73*(4), 357–371. https://doi.org/10.1016/j.jenvman.2004.07.010

Hirsch, P. D., Adams, W. M., Brosius, J. P., Zia, A., Bariola, N., & Dammert, J. L. (2010). Acknowledging conservation trade-offs and embracing complexity. *Conservation Biology, 25*(2), 259–264. https://doi.org/10.1111/j.1523-1739.2010.01608.x

Holden, M. (2001). Uses and abuses of urban sustainability indicator studies. *Canadian Journal of Urban Research, 10*(2), 217–236.

Holden, M. (2013). Sustainability indicator systems within urban governance: Usability analysis of Sustainability Indicator Systems as boundary objects. *Ecological Indicators, 32,* 89–96. https://doi.org/10.1016/j.ecolind.2013.03.007

Hukkinen, J. (2004). Sustainability indicators for anticipating the fickleness of human-environmental interaction. In S. K. Sikdar, P. Glavič, & R. Jain (Eds.), *Technological choices for sustainability* (pp. 317–333). Springer, Berlin, Heidelberg. https://doi.org/10.1007/978-3-662-10270-1_20

Hák, T., Janoušková, S., Moldan, B., & Dahl, A. L. (2018). Closing the sustainability gap. *Ecological Indicators, 87,* 193–195. https://doi.org/10.1016/j.ecolind.2017.12.017

Jacques, P. (2021). *Sustainability: The basics.* Routledge.

Jasanoff, S. D. (2004). *States of knowledge. The co-production of science and social order.* Routledge.

Kay, J. J. (2008). Framing the situation. In D. Waltner-Toews, J. J. Kay & N. Lister (Eds.), *The ecosystem approach* (pp. 15–34). Columbia University Press.

Kourantidou, M., Hoover, C., & Bailey, M. (2020). Conceptualizing indicators as boundary objects integrating Inuit knowledge and western science for Marine resource management. *Arctic Science, 6*(3), 279–306. https://doi.org/10.1139/as-2019-0013

Kroll, C., Warchold, A., & Pradhan, P. (2019). Sustainable development goals (SDGs): Are we successful in turning trade-offs into synergies? *Palgrave Communications, 5*(1), 140. https://doi.org/10.1057/s41599-019-0335-5

Latour, B. (1993). *We have never been modern.* Harvard University Press.

Lehtonen, M., Sébastien, L., & Bauler, T. (2016). The multiple roles of sustainability indicators in informational governance: Between intended use and unanticipated influence. *Current Opinion in Environmental Sustainability, 18,* 1–9. https://doi.org/10.1016/j.cosust.2015.05.009

Li, M., Wiedmann, T., Fang, K., & Hadjikakou, M. (2021). The role of planetary boundaries in assessing absolute environmental sustainability across scales. *Environment International, 152,* 106475. https://doi.org/10.1016/j.envint.2021.106475

Lloyd, E. A. (2012). The role of 'complex' empiricism in the debates about satellite data and climate models. *Studies in History and Philosophy of Science Part A, 43*(2), 390–401. https://doi.org/10.1016/j.shpsa.2012.02.001

Longino, H. E. (1990). *Science as social knowledge: Values and objectivity in scientific inquiry.* Princeton University Press.

Marr, D. (1982). *Vision.* Freeman.

Mascarenhas, A., Nunes, L. M., & Ramos, T. B. (2015). Selection of sustainability indicators for planning: Combining stakeholders' participation and data reduction techniques. *Journal of Cleaner Production, 92,* 295–307. https://doi.org/10.1016/j.jclepro.2015.01.005

Meadows, D. (1998). *Indicators and information systems for sustainable development.* The Sustainability Institute. https://donellameadows.org/archives/indicators-and-information-systems-for-sustainable-development/.

Merino-Saum, A., Halla, P., Superti, V., Boesch, A., & Binder, C. R. (2020). Indicators for urban sustainability: Key lessons from a systematic analysis of 67 measurement initiatives. *Ecological Indicators, 119,* 106879. https://doi.org/10.1016/j.ecolind.2020.106879

Miller, C. A. (2005). New civic epistemologies of quantification: Making sense of indicators of local and global sustainability. *Science, Technology, & Human Values, 30*(3), 403–432. https://doi.org/10.1177/0162243904273448

Miller, T. R. (2019). Imaginaries of sustainability: The techno-politics of Smart Cities. *Science as Culture, 29*(3), 365–387. https://doi.org/10.1080/09505431.2019.1705273

Mol, B. (2006). Environmental governance in the Information Age: The emergence of informational governance. *Environment and Planning C: Government and Policy, 24,* 497–514. https://doi.org/10.1068/c0508j

Molden, B. and A. Dahl (2007). Challenges to sustainability indicators. In H. Tomás, B. Moldan & A. Dahl (Eds.), *Sustainability indicators* (pp. 1–24). Island Press.

Mooney, C., Eilperin, J., Butler, D., Muyskens, J., Narayanswamy, A., & Ahmed, N. (2021, November 8). 'Countries' climate pledges built on flawed data, Post investigation finds'. *Washington Post.* https://www.washingtonpost.com/climate-environment/interactive/2021/greenhouse-gas-emissions-pledges-data/

Morse, S. (2004). *Indices and indicators in development.* Routledge.

Muhl, E. K., Armitage, D., Silver, J., Swerdfager, T., & Thorpe, H. (2022). Indicators are relational: Navigating knowledge and power in the development and implementation of coastal-marine indicators. *Environmental Management, 70*(3), 448–463. https://doi.org/10.1007/s00267-022-01670-3

Murray, S., Galik, C. Bast, J., & Hawley, D. (2022). *From "think" to "do": Operationalizing the sustainable development goals in university curricula.* U.N. Chronicle. https://www.un.org/en/un-chronicle/think-do-operationalizing-sustainable-development-goals-university-curricula

Murray, T., Waltner-Toews, D., Sanchez-Choy, J., & Sanchez-Savala, F. (2008). Food, floods, and farming: An ecosystem approach to human health on the Peruvian amazon frontier. In D. Waltner-Toews, J. J. Kay & N. Lister (Eds.), *The ecosystem approach* (pp. 213–235). Columbia University Press.

Norton, B. G. (2005). *Sustainability: A philosophy of adaptive ecosystem management.* University of Chicago Press.

Norton, B. G. (2015). *Sustainable values, sustainable change.* University of Chicago Press.

Oberlack, C., Sietz, D., Bürgi Bonanomi, E., de Bremond, A., Dell'Angelo, J., Eisenack, K., Ellis, E. C., Epstein, G., Giger, M., Heinimann, A., Kimmich, C., Kok, M. T., Manuel-Navarrete, D., Messerli, P., Meyfroidt, P., Václavík, T., & Villamayor-Tomas,

S. (2019). Archetype analysis in sustainability research: Meanings, motivations, and evidence-based policy making. *Ecology and Society*, *24*(2), 26. https://doi.org/10.5751/es-10747-240226

O'Neill, J., Holland, A., & Light, A. (2008). *Environmental values*. Routledge.

Payakkamas, P., de Kraker, J., & Dijk, M. (2023). Transformation of the urban energy–mobility Nexus: Implications for sustainability and equity. *Sustainability*, *15*(2), 1328. https://doi.org/10.3390/su15021328

Persson, J., Johansson, E. L., & Olsson, L. (2018). Harnessing local knowledge for scientific knowledge production: Challenges and pitfalls within evidence-based sustainability studies. *Ecology and Society*, *23*(4), 38. https://doi.org/10.5751/es-10608-230438

Pintér, L., Hardi, P., Martinuzzi, A., & Hall, J. (2012). Bellagio stamp: Principles for sustainability assessment and measurement. *Ecological Indicators*, *17*, 20–28. https://doi.org/10.1016/j.ecolind.2011.07.001

Porter, T. (1996). *Trust in numbers*. Princeton University Press.

Pradhan, P., Costa, L., Rybski, D., Lucht, W., & Kropp, J. P. (2017). A systematic study of sustainable development goal (SDG) interactions. *Earth's Future*, *5*(11), 1169–1179. https://doi.org/10.1002/2017ef000632

Pupphachai, U., & Zuidema, C. (2017). Sustainability indicators: A tool to generate learning and adaptation in Sustainable Urban Development. *Ecological Indicators*, *72*, 784–793. https://doi.org/10.1016/j.ecolind.2016.09.016

Ramsey, J. (2017). Something wicked this way comes. *Ethics, Policy & Environment*, *20*(3), 247–250. https://doi.org/10.1080/21550085.2017.1374047

Randall, A. (2021). Monitoring sustainability and targeting interventions: Indicators, planetary boundaries, benefits and costs. *Sustainability*, *13*(6), 3181. https://doi.org/10.3390/su13063181

Rydin, Y., Holman, N., & Wolff, E. (2003). Local sustainability indicators. *Local Environment*, *8*(6), 1–1. https://doi.org/10.1080/762742058

Sari, D. A., Margules, C., Boedhihartono, A. K., & Sayer, J. (2018). Criteria and indicators to audit the performance of complex, multi-functional forest landscapes. In S. Bell & S. Morse (Eds.), *Routledge handbook of sustainability indicators* (pp. 407–426). Routledge.

Schoen, D., & Rein, M. (1993). *Frame reflection: Towards the resolution of intractable policy conflicts*. Basic Books.

Scott, J. (1999). *Seeing like a state*. Yale University Press.

Sen, G., & Mukherjee, A. (2014). No empowerment without rights, no rights without politics: Gender-equality, mdgs and the post-2015 development agenda. *Journal of Human Development and Capabilities*, *15*(2–3), 188–202. https://doi.org/10.1080/19452829.2014.884057

Seto, K. C., Davis, S. J., Mitchell, R. B., Stokes, E. C., Unruh, G., & Ürge-Vorsatz, D. (2016). Carbon lock-in: Types, causes, and policy implications. *Annual Review of Environment and Resources*, *41*(1), 425–452. https://doi.org/10.1146/annurev-environ-110615-085934

Shrader-Frechette, K. (2013). [Review of the book *Unsimple Truths*, by S. Mitchell]. *The British Journal for the Philosophy of Science*, *64*(2), 449-453. https://doi.org/10.1093/bjps/axs023

Stanners, D., Bosch, P., Dom, A., Gabrielsen, P., Gee, D., Martin, J., ... & Weber, J. L. (2007). Frameworks for environmental assessment and indicators at the EEA. In

T. Hák, B. Moldan & A. Dahl (Eds.), *Sustainability indicators* (pp. 127–144). Island Press.

Star, S. L. (2010). This is not a boundary object: Reflections on the origin of a concept. *Science, Technology, & Human Values, 35(5)*, 601–617. https://doi.org/10.1177/0162243910377624

Star, S. L., & Griesemer, J. R. (1989). Institutional ecology, 'translations' and boundary objects: Amateurs and professionals in Berkeley's Museum of Vertebrate Zoology, 1907–39. *Social Studies of Science, 19*(3), 387–420. https://doi.org/10.1177/030631289019003001

Steffen, W., Richardson, K., Rockström, J., Cornell, S. E., Fetzer, I., Bennett, E. M., Biggs, R., Carpenter, S. R., de Vries, W., de Wit, C. A., Folke, C., Gerten, D., Heinke, J., Mace, G. M., Persson, L. M., Ramanathan, V., Reyers, B., & Sörlin, S. (2015). Planetary boundaries: Guiding human development on a changing planet. *Science, 347*(6223), 1259855. https://doi.org/10.1126/science.1259855

Stiglitz, J. E. (2009). GDP fetishism. *The Economists' Voice, 6*(8), 1–3. https://doi.org/10.2202/1553-3832.1651

Stockholm Resilience Centre. (n.d.) *Planetary boundaries.* https://www.stockholmresilience.org/research/planetary-boundaries.html

Suppes, P. (1962). Models of data. In E. Nagel, P. Suppes & A Tarski (Eds.), *Logic, methodology, and philosophy of science: Proceedings of the 1960 international congress* (pp. 252–261). Stanford University Press.

Sébastien, L., & Bauler, T. (2013). Use and influence of composite indicators for sustainable development at the EU-level. *Ecological Indicators, 35*, 3–12. https://doi.org/10.1016/j.ecolind.2013.04.014

Sébastien, L., Bauler, T., & Lehtonen, M. (2014). Can indicators bridge the gap between science and policy? An exploration into the (non)use and (non)influence of indicators in EU and UK policy making. *Nature and Culture, 9*(3), 316–343. https://doi.org/10.3167/nc.2014.090305

Termorshuizen, J. W., & Opdam, P. (2009). Landscape services as a bridge between landscape ecology and Sustainable Development. *Landscape Ecology, 24*(8), 1037–1052. https://doi.org/10.1007/s10980-008-9314-8

Tscherning, K., Helming, K., Krippner, B., Sieber, S., & Paloma, S. G. (2012). Does research applying the DPSIR Framework Support Decision Making? *Land Use Policy, 29*(1), 102–110. https://doi.org/10.1016/j.landusepol.2011.05.009

Turnhout, E., Hisschemöller, M., & Eijsackers, H. (2007). Ecological indicators: Between the two fires of science and policy. *Ecological Indicators, 7*(2), 215–228. https://doi.org/10.1016/j.ecolind.2005.12.003

UNDP (n.d.). *Human development index.* United Nations Development Programme. https://hdr.undp.org/data-center/human-development-index#/indicies/HDI

UNEP (1997). *Global environment outlook.* Oxford University Press.

van Fraassen, B. (2008). *Scientific representation: Paradoxes of perspective.* Oxford University Press.

Verma, P., & Raghubanshi, A. S. (2018). Urban sustainability indicators: Challenges and opportunities. *Ecological Indicators, 93*, 282–291. https://doi.org/10.1016/j.ecolind.2018.05.007

Waas, T., Hugé, J., Block, T., Wright, T., Benitez-Capistros, F., & Verbruggen, A. (2014). Sustainability assessment and indicators: Tools in a decision-making strategy for sustainable development. *Sustainability, 6*(9), 5512–5534. https://doi.org/10.3390/su6095512

Wellstead, A., Howlett, M., & Rayner, J. (2016). Structural-functionalism redux: Adaptation to climate change and the challenge of a science-driven policy agenda. *Critical Policy Studies, 11*(4), 391–410. https://doi.org/10.1080/19460171.2016.1166972

Wesselink, A., Buchanan, K. S., Georgiadou, Y., & Turnhout, E. (2013). Technical knowledge, discursive spaces and politics at the science–policy interface. *Environmental Science & Policy, 30*, 1–9. https://doi.org/10.1016/j.envsci.2012.12.008

Wesselink, A., & Warner, J. (2010). Reframing floods: Proposals and politics. *Nature and Culture, 5*(1), 1–14. https://doi.org/10.3167/nc.2010.050101

Winner, L. (1980). Do artifacts have politics? *Daedalus, 109,* 121–136.

Winsberg, E. (2018). *Philosophy and climate science.* Cambridge University Press.

Wu, J., & Wu, T. (2012). Sustainability indicators and indices: An overview. In C. Madu & C. Kuei (Eds.), *Handbook of sustainability management* (pp. 65–86). Imperial College Press.

# 5   Ethics and Sustainability

We try to discover the meaning of sustainability, theorize about it, and get evidence pertaining to it because it is a question about social and natural well-being. It "is an inherently normative concept, rooted in real world problems and very different sets of values and moral judgments" (Robinson, 2004, p. 379; cf. Curren & Metzger, 2017b; Miller et al., 2013; Norton, 2005, 2015). How should we understand this normativity? In Chapter 3 I argued the folk view of scientific theories, in which claims are derived from or viewed as instantiations of law-like, universally valid generalizations, masks several theoretical strategies that are useful in the pursuit of sustainability. Here, I argue that use of an analogous approach to moral normativity leads to difficulties. Many sustainability proposals employ a "theoretical-juridical" (Walker, 2007) conception of ethical decision-making, in which claims about the rightness or wrongness of actions are seen to follow from logically prior assumptions about fundamental moral knowledge. However, this model

> conceal[s] the complex way that norms of human action both shape and are shaped by the natural world. . . [It] forces conversations about sustainability into an awkward vocabulary incapable of expressing the sense in which sustainability is a moral ideal.
>
> (Thompson, 2010, p. 2)

It assumes we already know what well-being means in the messy, partially understood matters of sustainability. In so doing, it elides issues associated with place, power, language, and gender. It downplays the existence and therefore the importance of trade-offs, path dependencies and policy resistances, sites where we often discover and justify our obligations. And it carves out roles for both moral and scientific knowledge that reinforce a knowledge-first, expert-based approach to sustainability.

Margaret Urban Walker's (2007) 'expressive-collaborative' model of ethical decision-making offers an alternative approach to understanding sustainability as a moral ideal and practice. Communities begin deliberation already in situations of concern, working from within to criticize and extend practices

DOI: 10.4324/9781003268697-5

to understand responsibilities and obligations. Every criticism and extension must be justified. On this model, wicked sustainability problems become sites of contestation where norms regarding the pursuit of sustainability are discovered and justified. This model does not deliver moral judgments directly, providing instead questions and concerns that must be addressed if a claim is to be considered ethically justified.

After characterizing the theoretical-juridical model of ethical decision-making, I argue that it is present in many sustainability proposals. The examples represent a range of traditional ethical approaches: utilitarianism, Kantianism, pragmatism and discourse ethics, and virtue theory. All produce awkwardness, albeit of different substantive sorts, about what the moral ideals of sustainability are. I then sketch the expressive-deliberative model and gesture toward some examples in the sustainability literature that I think illustrate it. As in previous chapters, the message is to look to context and practices as we justify actions as being more sustainable.

## The Theoretical-Juridical Conception of Ethics and Morals

In a theoretical-juridical model of ethical decision-making, morality is represented as a theory and as a jury or judge. Morality is represented as

> a surprisingly compact kind of theory or some kind of internal guidance system of an agent that could be modeled by that kind of theory. It makes morality look as if it consists in, or could be represented by, a compact cluster of beliefs.
>
> (Walker, 2007, p. 8)

Typically, the beliefs are conceived as expressing law-like generalizations. The generalizations are more primitive, logically prior to specific moral judgments. If we want to make a judgment, we deduce or instantiate it from the generalizations. The model is juridical because the theories are "seen as delivering or justifying verdicts on cases (jury or judge, as it were)" (Walker, 2007, p. 43; cf. Allhoff, 2010; Valera et al., 2020).[1] Utilitarian, contractarian, Kantian (and neo-Kantian) and rights-based theories all "realize or approximate" this model (Walker, 2007, p. 8).

This picture of moral inquiry is intellectualist, rationalist, individualist, modular, and transcendent. It seats morality in some central, specifically moral beliefs. It assumes that we understand and test those beliefs "by reflection on concepts and logical analysis of the relations of evidential support among moral beliefs" or "on the 'deliverances' of intuitions" (Walker, 2007, p. 9). The beliefs equip individual agents with a guidance system. All agents are fungible, i.e., the moral beliefs tell one what is right to do no matter who one is or what kind of life one might be living. This gives rise to its modularity:

we simply apply the same moral module to all the different situations we find ourselves in. It is transcendent, the core being applicable to all cultures, histories, and material conditions.

## Theoretical-Juridical Morals and Ethics in Sustainability

### *The Brundtland Definition of Sustainability*

The Brundtland Report defines sustainable development as "development that meets the needs of the present without compromising the ability of future generations to meet their own needs" (WCED, 1987). As O'Neill et al. (2008, pp. 183–201) argue, the definition – as it is interpreted in much of the economic development literature and in many government agencies – falls squarely within the welfarist interpretation of utilitarianism. Human welfare is the 'what' of sustainability. It is to be sustained for present and future humans. It is sustained in order to maximize welfare over time. Sustainability becomes a matter of maintaining well-being by maximizing welfare, with 'welfare' further interpreted as preference satisfaction. Preferences can be for anything. And they are fungible: if preferences are satisfied, it does not matter what satisfies them. Satisfying preferences requires input, usually conceptualized as 'capital.' A distinction is usually made between natural and human capital. Natural capital includes organic and inorganic resources, broadly conceived. Human capital includes machines, roads, knowledge, and skills. Further distinctions between 'renewable' and 'nonrenewable' capital of various sorts are often made.

Utilitarianism and its mainstream welfare economics incarnation are prime examples of theoretical-juridical thinking. Morality is determined by beliefs about well-being, interpreted broadly as utility and more specifically as preference satisfaction. It is rationalist because right action is discovered by calculating which consequences maximize the well-being of affected agents. (Admittedly, many utilitarians recognize limitations here, but the usual response is to try to find ways around the problem by finding proxies for any ill-understood consequences.) It is individualist, and agents are fungible. And it is transcendent, applicable to all people and situations.

Criticisms of utilitarianism, mainstream welfare economics, and the Brundtland definition abound. I use examples from O'Neill et al. (2008, pp. 183–201) to illustrate the awkwardness produced when the model is applied to matters of sustainability. Since utilitarian conceptualizations of sustainability require capital input to maintain welfare, to what extent are different forms of capital substitutable? It seems some items of natural capital are conditions that are necessary for life, much less a flourishing one. More entertainment or better housing will not help a person suffering from malnutrition. Even if one is not in danger of losing one's life, it seems some items cannot be substituted for each other. The life history of a community in a particular place may make it reasonable for the community to reject any compensation for giving up the physical location. Money is not substitutable for place. When utilitarianism, especially in its welfare economics incarnation, assumes different kinds of goods are substitutable for each other, it ignores independent dimensions of human well-being. That is, it avoids questions about how human welfare is shaped in incommensurable ways by the natural world.

### *Living Well Now and In the Future: Why Sustainability Matters,* Randall Curren and Ellen Metzger

Curren and Metzger (2017b) develop an ethic of sustainability that is broadly Kantian, with elements of virtue ethics, Aristotelian *eudaimonia*, and Rawlsian justice as fairness playing supporting roles. Their ethical theory is "an application of principles of common morality informed by the sustainability facts of life" (p. 54). The first principle of common morality is respect for persons as rational beings. This grounds our duties. Given that every person is a moral agent and that everyone has ends they are trying to fulfill, "respectful self-restraint and cooperation" are norms that "it is rational for us all to adopt . ... Respectful self-restraint entails taking care not to cause harm, and fulfilling this duty of care requires attention" to make sure one's conduct, both actually and potentially, is not harmful (Curren & Metzger, 2017b, p. 56). The duty of care is the second principle of common morality.

The sustainability facts of life are ecological, throughput and socio-political sustainability. Ecological (un)sustainability refers to the (in)stability of natural systems when used by some group of humans. Throughput (un) sustainability refers to whether human practices use a system's provisions (water, land productivity, etc.) in ways that do or do not allow the system to regenerate that provision. The two notions are linked but separate: societies can draw down the provisioning capacity for some time before any ecological instability is produced. Socio-political (un)sustainability is characterized as the collapse or durability of social and political practices and institutions.

The principles of common morality and the sustainability facts of life are the basis of five sustainability principles. Delineating how they lead to the first principle is sufficient for my purposes. "Take care to ensure that the totality of human practices is ecologically sustainable" is "a straightforward implication of a basic duty to take care not to harm" (Curren & Metzger, 2017b, p. 58). Renewable natural capital (RNC) is the sustainability fact of life that informs this principle. RNC is a natural asset that yields goods or services and is active and self-maintaining.

Certain kinds of people and institutions are needed to employ this principle (and the other four). People are characterized *via* virtue ethics. Virtues allow people to act well. The virtues of sustainability include wisdom, justice, moderation, and courage. To characterize the institutions, they accept Rawl's notion of justice as fairness but modify it so that it includes discussion of how societal practices and institutions alter the structure of opportunity for people on different social levels. The aim of institutions is to enable all members of a society to live well. Their conception of 'living well' is "inspired by the idea of *eudaimonia* . . . in Aristotle's works" (Curren & Metzger, 2017b, p. 80). For individuals, "living well (*eudaimonia*) is largely a product of fulfilling human potential in positive ways" (Curren & Metzger, 2017b, p. 71). Institutions "exist *to enable all of its members to live well* and should provide opportunities *sufficient* to enable all to do so" (Curren & Metzger, 2017b, p. 80, emphasis in original). Basic Psychological Needs Theory is used to characterize positive ways of fulfilling potential. To live well in positive ways, one needs: autonomy, i.e., the belief that one can choose one's own behaviors and actions; competency, i.e., the ability to work effectively; and relatedness, i.e., the experience of forming strong bonds with others.

The account is clearly modeled along theoretical-juridical lines. Deductivism is present throughout: the ethic of respect 'grounds' our duties; respectful self-restraint 'entails' a duty of care; and the first sustainability principle is 'a straightforward implication' of the duty of care. Likewise, the principles "*require* societies" to observe ecological, throughput and sociopolitical sustainability" (Curren & Metzger, 2017b, p. 85, emphasis added). Similar language is present throughout the book. It is also rationalist: moral knowledge, based in reflection on what is rational, is logically prior to action. The account is explicitly universalist and transcendent: the "conceptualization of *universal necessities* for a good life offers a *suitably timeless account*" (Curren & Metzger, 2017b, p. 85, emphasis added) of people's needs and the institutional arrangements that will allow people to live well.

I think the framework pushes conversations about sustainability into an awkward vocabulary in several ways. First, consider Curren and Metzger's use of ancient philosophy. They read Plato's *Republic* as a dialog about overconsumption. The luxurious "city with a fever" of Book 2 contrasts with the "true" and "healthy" city, which is "sustainable across generations" and in which there is no greed and injustice (Curren & Metzger, 2017b, p. 70). They omit the fact that Plato abandons this simplest of cities because justice and injustice cannot be located there. Also, in Annas' (1981) prominent interpretation, Plato is not interested in the simple city *per se*. Rather, the example introduces Plato's idea that each person is naturally suited to a particular role in society, and that each person should play that role. Plato's simple city seems far removed from concerns about sustainability and justice.

Likewise, their appeal to *eudaimonia* seems only remotely related to Aristotle. They replace Aristotle's concern with whole lives with lives lived considering the sustainability facts of life. Patterns of education, governance, etc., flow from the conception of sustainability. However, for Aristotle being virtuous could involve muting or overriding concerns about sustainability in favor of other concerns that make a whole life virtuous. Everything depends on the person (or society) and the situation.[2] To be fair, Curren and Metzger (2017b) do say only that "Aristotle's moral thought . . . is an important point of departure" (p. 71) for their account and that their account is "inspired" (p. 80) by Aristotle. Nonetheless, the relations need to be spelled out more clearly in order to avoid the implication that the Aristotle provides direct support for their framework.

The more fully Aristotelian take on *eudaimonia* is present in sustainability discussions. Efficacy may involve privileging, for some length of time, other dimensions of lives at the expense of ecological, throughput or socio-political sustainability (Jupiter, 2017; McCarter et al., 2018). Often, a concern for livelihoods drives that discussion (Cruz-Torres & McElwee, 2012; McShane et al., 2011; McShane & Newby, 2004; Sarkar & Montoya, 2011). In other cases, being able to satisfy a bundle of basic needs competently may be thwarted by the presence of trade-offs (Schneider, 1997). Additionally, it is recognized that efficacy at the individual level may be at odds with efficacy at the group level (Geels, 2012; Shove, 2010). And so on. Overall, consideration of place-based factors – and the attendant questions about trade-offs, acceptable policy resistances and tolerable path dependencies – is often far more important than an appeal to general notions of sustainability.

Second, consider the philosophy of action embedded in the framework. We can remove the systemic barriers that prevent us from living sustainably if we change our "human attributes, practices, norms, settings, structures, cultures, institutions, systems and policies," understand "what is conducive to cooperation," and engage in "self-organized collective efforts" (Curren & Metzger, 2017a, pp. 231–232). In short, we adopt attitudes of respectful self-restraint and cooperation. Once we adopt them, we choose sustainable behaviors. Change follows. This is what Shove (2010) calls the 'ABC' model of social change: values and attitudes (A) are believed to drive the kinds of behavior (B) that autonomous individuals choose (C) to adopt freely. The model was developed for people, but it applies to any entity that is conceived as an individual in its larger context. Plausibly, the "structural-functionalist" view of policy-making (Wellstead et al., 2016) mentioned in the previous chapter is an institutional-level version of the ABC model.

There are good reasons to question this philosophy of action. If we recognize that effects are "never in isolation and that interventions go on within, not outside, the processes they seek to shape," and that conventions (of washing clothes, washing people, traveling to work, etc.) are "themselves sustained and changed through the ongoing reproduction of social practice" (Shove, 2010, p. 1279), values and behaviors become linked in self-modifying, nonlinear systemic interaction (cf. Kurz et al., 2014). Individuals are not autonomous, free agents; rather, they are carriers of practices that have an inertia associated with them.

The view of individuals as carriers of practice makes sense in sustainability contexts. Practices of many sorts reinforce each other. Systems develop inertial resistance to change, a phenomenon known as 'lock-in' or 'path-dependency' (cf. Seto et al., 2016). Since the system constrains your options, it is not enough to think you can freely change your attitudes and behaviors. This point applies directly to policy makers. They are not autonomous agents. Legal limitations, funding mechanics, lack of access to research and data, and the culture of government all make lock-in a real feature of the institutions involved in setting environmental policy (National Research Council, 2013).[3] Path-dependencies are associated with scalar effects. Actions to preserve opportunities at one scale may eliminate opportunities at another. Driving electric vehicles, which seems the right thing to do given ecological and throughput sustainability (and probably socio-political sustainability), often leads to communities driving more miles (Chakraborty et al., 2021). If the electricity for battery-powered cars comes from the burning of fossil fuels, which is common in many places around the globe (Mitchell, 2022), driving more miles in an electric vehicle remains problematic from a sustainability point of view. If the electricity is renewable but generated from sites that are near endangered species habitats, on steep gradients, or on less well-regulated private lands, questions about environmental impacts of renewable energy emerge (Parmar et al., 2023). Consider also the need to maintain existing roads and build new ones as more miles are being driven. Suddenly, simply choosing to engage in an action based on a change in attitude seems less straightforward. Policy resistances rear their ugly head. It is also worth keeping in mind that as you solve your 'systemic action problem,' the system may move to a new configuration, thwarting your attempt to solve the original problem (Mayumi & Giampietro, 2006). Driving more miles in electric cars could exacerbate gridlock.

Third, the framework reinforces some aspects of the traditional relation between science and society. Sustainability topics and actions are chosen in a conversation between the public and the experts. Knowledge about those matters are produced by the experts (Curren & Metzger, 2017b, pp. 93–94). This largely reinforces the separation of science and society, allowing citizens a role only in specifying which problems will be addressed. Questions about what the problem is and who it is a problem for are not directly discussed. Having someone tell you what the problems are can be deeply at odds with sustainability, not least since it interferes with your autonomy.

### *Bryan Norton – Pragmatism, Proceduralism, Discourse Ethics, and Community Values*

Bryan Norton uses 'sustainability' to name his general approach to environmental ethics. As outlined in Chapter 3, he proposes that effective action toward normative sustainability can be achieved if a community articulates and adopts a set of community values on which they have deliberated reflectively and democratically. Since the community is concerned about its future citizens, it must have in view the stuff that allows its members to live well in the future. Norton's version of Curren and Metzger's "sustainability

facts of life" is a commitment to "normative sustainability." Attention to resilience (as a normative as well as a descriptive term), self-organization, and
hierarchical organization allow a community to understand what elements in
our environments are "essential to sustainable living" (Norton, 2015, p. 94).

For Norton, actions are justifiable insofar as a community value consensus
is elicited and constructed in fair and effective decision procedures. To do this,
a community must commit itself to procedural rationality in an ideal community of reason givers and criticizers. "In an appropriate, democratic public
discourse, all opinions and all opinion-givers deserve respect, in the sense that
they are responded to with reasons and evidence, not force or ridicule" (Norton, 2015, p. 168). Disagreements are "aired and discussed by giving reasons,
rather than by name calling or use of power" (Norton, 2015, p. 168). Good
deliberation "requires the existence of a common language in which their discourse takes place" (Norton, 2015, p. 168). For a community to use adaptive
management successfully and collaboratively, this full process is needed.

The deliberative process identifies and uses values. Values of various sorts
offer "a sort of roadmap toward a definition of sustainable living for a community," suggesting "the underlying logical structure of the sustainability idea" and
helping "communities to ask the right questions in their adaptive management
processes" (Norton, 2005, p. 359). Community-identity values are particularly
important. These "are obligations that thoughtful community members, committed to cooperative action and loving their place, would articulate and impose upon
themselves for the good of their community" (Norton, 2005, p. 372). Values are
not just believed; they are performed in acts that "signal a commitment to perpetuate stuff deemed of great value to the community, provided the costs are bearable"
(Norton, 2005, p. 372). When the community acts like an ideal speech community, "'thick' obligations . . . accepted as a part of defining and living a good life in
a community with shared values" are identified (Norton, 2005, p. 372).

When a community embraces proceduralism and has identified its values,
the framework offers a powerful way to enact a thickly evaluative notion of
sustainability. Norton (2015, pp. 218–257) discusses two communities that
have proceeded in this way. The communities recognized the wicked nature of
sustainability problems, used language that was inseparably descriptive and
prescriptive, gave reasons chaotically, and articulated thick obligations. Given
this, it appears Norton has avoided the theoretical-juridical model of ethical
decision-making.

However, the theoretical-juridical model asserts itself at the meta-level.
The commitment to procedural rationality in an ideal community of reason
givers and criticizers is based in Habermasian discourse ethics (Norton, 2005,
pp. 278–289). This approach is transcendent, individualist, modular and rationalist, with the latter including evidential support from descriptive, prescriptive, and mixed claims. The

core procedural rules of fairness and unbiased treatment of all participants are based on a universal foundation and apply to all who engage

in deliberative discourse; they exist independently of culture and they hold quite independently of the particular beliefs and values of particular participants.

(Norton, 2005, p. 289)

There is nothing wrong with this as an ideal, but deviations occur. Focusing on them provides insight into how Norton's proceduralism elides attention to contextual factors that are relevant to the process of articulating sustainability. As a method, proceduralism can produce surface-level agreement. "[P]rocedural requirements may be less ethically problematic. Justifications of procedural requirements can elide at least some conflicts among basic values" (Battin et al., 2009, p. 315). Norton (2015, p. 77) seems to recognize this, adding a substantive requirement of the absolute necessity of maintaining natural capital. However, he then argues that the question about which kinds of natural capital to preserve will be addressed in the process of eliciting and constructing the consensus. But again, if the process – as process – elides some of the conflicts among basic values, the agreement about natural capital may be surface-level only.

Appeals to common language and public reason-giving raise questions of epistemic injustice, injustice "done to someone specifically in their capacity as a knower" (Fricker, 2007, p. 1). This is a commonly recognized problem regarding Habermasian discourse ethics (Allen, 2017; De Brasi & Warman, 2023; Fraser, 1990; Young, 1990). And it shows up in matters of sustainability. Seeking a common language for sustainability often allows communities to think they are including all perspectives when in fact the language devalues some speakers and knowledge (Domínguez & Luoma, 2020; Kashwan et al., 2021). Indicators are a case in point. As I noted in the previous chapter, seeing indicators as providing a common language (a view that Norton endorses) often devalues kinds of knowledge not expressed by the indicators, erases assumptions involved in their production, and depoliticizes knowledge. Additionally, use of a common language can be traumatic and dangerous in some situations. For Iñupiaq or Yupik communities in Alaska, where policies of cultural erasure mandated use of English as a common language, use of English is a form of trauma for the elders (Reo et al., 2019). Use of a common language also increases risks on the dangerous Alaskan coast. "Hunting and gathering parties who have to use strictly English or a blended heritage-language-English may be less well equipped to deal with risks because their communication is slowed, and because English is experienced as a less direct language" (Reo et al., 2019, p. 222).

The ideal of open, public reason-giving about opinions (with or without a common language) ignores language as a form of symbolic power. "[R]eason-ableness is itself a social construction which usually benefits those already in power. ... [A]ppeals to intersubjectively valid norms can be a way of structuring contestation to perpetuate the status quo" (Kohn, 2000, p. 408; cf. Arney et al., 2022; Kothari, 2001; Merino-Saum, 2018; Williams, 2004). Other forms of discourse such as greetings, rhetoric, and narratives are often essential to inclusive deliberation (Young, 2000). In some Native American tribes, these are central to learning and being a community member (Simpson, 2017). Storytelling disrupts power relations in society (Sze, 2020) and is often central to working out what sustainability means for a community (Whyte et al., 2018).[4] More than public reason giving and response is needed.

Thinking that a community has a single identity can be problematic. When mobilizing against the expansion of a US military base in 2010, the inhabitants of Guahan/Guam mobilized both their identities as patriotic supporters of the United States and as colonized people who have been treated unjustly by the United States (Bevacqua & Bowman, 2018). In some instances, community identities are so complex that it is impossible to identify *a* community. In Melanesia, communities can be comprised of deeply intertwined land-kinship groups, with different groups claiming rights to primary and secondary access and control over a *puava*, a delimited territory of land, sea, and/or reefs. In one instance, it proved impossible to create a uniform version of descent groups and landownership for negotiation over leases with mining representatives (Jupiter, 2017). If these examples seem too exotic, think of the multiple identities expressed even when a community is addressing a 'common' problem.

Finally, the procedural model – at least as Norton uses it – assumes the ABC model of social change. When a community is confronted by a problem, community members enter a deliberative phase. They elicit and then construct a community value consensus, consider what should count as a rule of discourse, and suggest new language. Experimental action (since this occurs within an adaptive management paradigm) then takes place in an "action phase" (Norton, 2005, p. 280). That is, discussion of values suggests the underlying structure of the sustainability idea, and then discussion and reflection leads to changes in behavior that the community chooses to enact, however experimentally. The point raised earlier is relevant. Since the system constrains actions, a community cannot freely change all attitudes and behaviors however deeply it deliberates and identifies values that imply change. Behaviors are tied to larger structural features of systems. I do not mean to suggest that change is impossible, but community choices do not float freely above existing social and physical processes and practices.

### Paul Thompson – Virtue Ethics

In *The Agrarian Vision,* Paul Thompson uses virtue ethics to articulate the moral ideals of sustainability. The question about what action is right can be answered only from within situations that are

> rich in relationships and responsibilities. . . . A good person is someone who recognizes and discharges the particular duties of his or her particular

station in life, realizing him- or herself through the lifelong formation and expression of a virtuous moral character.

(Thompson, 2010, p. 55)

Thompson articulates three ideals of virtue associated with agrarianism specifically and sustainability more generally. The Central Agrarian Tenet (CAT) states that farmers make the best citizens. Because their wealth is tied to the interests of the state, farmers care directly about the state and are motivated to participate in its governance. The Principle of Agrarian Localism (PAL) highlights the fact that farmers are dependent on local, not generalized, others. For those who are not farmers, localism provides the basis for understanding one's ties to the broader community and why they are needed. The Great Agrarian Goal (GAG) highlights self-realization as "the driving force behind all forms of voluntary human action" (Thompson, 2010, p. 55). GAG looks to the family farm as the place where individuals and families can most reliably develop the classical virtues of "solidarity, self-reliance, stewardship, faith, hope, and charity" (Thompson, 2010, p. 191).

Thompson's version of Curren and Metzger's 'sustainability facts of life' is "functional integrity." The stability and reproducibility of systems is critical to our ways of lives. Such concerns encompass biophysics, society, politics, psychology, and morality. We investigate whether our actions maintain the system or lead to its collapse.

For Thompson, the moral ideal of sustainability emerges when farmers express the agrarian virtues considering functional integrity. For those who do not farm or, like me, farm at a distance (both spatially and managerially), Thompson suggests participating in "focal practices." The practice that is the "culture of the table" connects us to food and farming. It includes being more deeply involved in the preparation and consumption of new and old foods in new ways. It means talking with and buying from local farmers and vendors. Buying locally connects you to cycles of production in your local place. This focal practice challenges the food system more fully and more consistently than vegetarianism, veganism, healthy eating, and ethical consumerism. As Thompson characterizes them, these practices often remain narrowly focused on asking if eating specific foods will make us healthier. In contrast, the agrarian ideals enlarge our focus to include virtuous individuals who have healthy bodies and are concerned with just communities and beautiful places.

The focus on specific relations to people, society and land and our embeddedness in all three clearly escapes the theoretical-juridical focus on universalism. But the way Thompson applies his version of virtue theory to matters of sustainability at times veers too closely to the theoretical-juridical model, leading to awkward takes on sustainability.

First, the combination of the three ideals and functional integrity can lead too quickly to claims about obligations. For instance, at one point Thompson (2010, p. 222) notes that sustainability presupposes "moral notions" of having sufficient food for the human population, of quality of life in rural communities, and fair wages in food production for rural workers. What if there are tensions between the production of sufficient food and the provision of fair wages? What if the production of sufficient food is in tension with ecological concerns? From some system standpoints, all agriculture involves such because it changes or freezes pre-existing species abundances and distributions. What if there are concerns about the ownership of that land, as there are in colonial contexts? Without more discussion about how the ideals often need to be balanced against each other, the conception starts to look thin.

Thompson also seems to not be concerned about different expressions of the agrarian ideals and their connection to the functional integrity of not-so-virtuous larger systems. His touchstones are Wendell Berry and Steinbeck's *Grapes of Wrath*, which he interprets as a song of sorrow about the loss of the agrarian ideal. Different pictures of life down on the farm are available. Watch *Oklahoma!*, with its many fights, distrust of outsiders, and Jud dying at the wedding festivities. Read Jane Smiley's *A Thousand Acres*, aka 'King Lear on the Farm.' If you want what someone once called "agrarian psychedelia," watch *Green Acres*. Or consider *Little House on the Prairie*. In the published versions, this is a classic example of a family managing the land and making it theirs. As Woodside (2016) notes, these themes were foregrounded in heavy edits by Laura Ingalls Wilder's daughter Rose Wilder Lane, with Ingalls Wilder agreeing to play up the theme of toughness in adversity. However, in Wilder's original memoirs the pioneers are stoic and often confused. A drought leads to the failure of the family's first homestead. They wander, working for a year in a tavern in Iowa. The experience is miserable, not least because the innkeepers murder some guests. A nine-month-old baby brother dies. A couple boards with the Wilders and is hardly stoic about the experience. The family moves from Iowa to Minnesota to North Dakota, homesteading on government grants each time.

At some level, the agrarian ideals are present in these examples. People own land, work the land, and are concerned with local governance. (In Rose Wilder Lane's telling, local governance for the Wilders is stereotypical libertarianism.) Characters are or become embedded in and deeply involved in community life (even if community life for some is largely family life). They are self-realizing, exhibiting many of the virtues associated with the Great Agrarian Goal. But farmers' desires to participate in local governance can go awry. Self-realization relies on the help of non-local others such as the government. Self-realizing individuals and communities express vices in concert with the agrarian virtues. Some of the instantiations are decidedly odd (and comical). In sum, rural life is far more complicated than indicated by appeal to the agrarian ideals and functional integrity.

Also, Thompson's account mostly elides issues of structural inequalities in society. For long periods of time in most agrarian societies, women had no right to property or the products of their labor. While Thompson recognizes that farming is not possible without the labor of women on a day-to-day basis, he sees the gender roles in farming as equally vital and complementary (Thompson, 2010, p. 189). This does not acknowledge directly the imbalance and unjustness of traditional roles (McKenna, 2011). To continue the arts theme, think of *Seven Brides for Seven Brothers*. The brothers are presented as almost paradigmatic examples of self-realizing individuals who learn – in truth they are taught by Milly, the first bride – how to govern themselves. The barn-raising scene draws directly from tropes picturing this as an activity involving essential dependence on local others. But it is worth remembering that although the story takes place in 1850 in progressive Oregon Territory, it would be seven years before the state constitution would guarantee women the right to own property without respect to marital status. In addition, one-sided gender roles and behaviors are woven into the story. The would-be brides are objects of competition in the barn dance. The brothers and the town beaus engage in fisticuffs with feet and then real fisticuffs (replete with dropping hammers on heads) to determine who gets to go a-courtin'. Sometime after the barn scene, the brothers kidnap (!) the would-be brides and cause an avalanche to stop the townspeople pursuing them. Thompson (2010, p. 85) recognizes that the agrarian ideals need to be reworked to avoid historical patterns of discrimination and domination, but this may not be as easy it seems. In agrarian and other societies today, women are still denied land rights (Kaaria & Osorio, 2018), and traditional roles continue to disadvantage women (McKenna, 2012). To reiterate a point from earlier, system structures inhibit the expression of individual behaviors, virtuous or not. As Thompson himself might say, the functional integrity of the system mitigates against change.

Similar comments can be made about race and class. Thompson recognizes but does not explore how such matters are shaped in agrarian ideals as they have been and are practiced. In the United States, freed slaves could put in a claim under the Homestead Act of 1862, but lack of resources, access only to poor quality land, and persecution by whites made it difficult to take advantage of the opportunity (Hayden et al., 2013). Native Americans were dispossessed by the Act. If we want to more adequately capture the moral ideals involved in sustainability, alleviating the structural inequalities present in agrarian and now industrial agricultural practices is called for.

### An Expressive-Collaborative Approach to Ethical Deliberation

What would a conception of ethical deliberation appropriate to sustainability as a moral ideal look like? As in other chapters, the approach I sketch does not provide a direct, substantive answer to the question. Rather, it elucidates the conditions that need to be met if we are to capture the responsibilities and obligations of sustainability. Theoretical-juridical approaches largely bypass this investigation, assuming as they do that the principles already provide a substantive characterization of morality and moral action.

Walker (2007) advances an "expressive-collaborative" approach to ethical deliberation. This has ties to MacIntyre (1981), Williams (1981), and Taylor (1989), but Walker explicitly avoids their totalizing conceptions of moral lives and projects (Walker, 2007, pp. 126–128, 137–159) and develops an account that has far more affinity with Addelson (1994). Questions of morality start with responsibilities and identities and how these are experienced and expressed in actual social life. We develop our senses of responsibility out of the narratives we construct about who and what we are related to. Those senses are "geographies of responsibilities, mapping the structure of standing assumptions that guides the distribution of responsibilities – how they are assigned, negotiated, deflected – in particular forms of moral life" (Walker, 2007, p. 105). Understanding this distribution means going beyond individuals and beyond specifically moral matters. Morality is reproduced between people, and it is about what is thought, perceived, felt, and acted out. This view of morality and moral knowledge is naturalized, critical and justificatory.

Natural moral knowledge identifies what kinds of things people need to know in order to act morally in their communities. Such knowledge allows us to know "for what and to whom we will have to account when we have done or failed to do something and what makes sense as a moral reason or excuse" (Walker, 2007, p. 68). It gives shape to reactive attitudes of blame, shame, etc. And it is reflexive, asking whether our understandings and attitudes deserve our authority. It tests whether what we feel is fitting.

Moral knowledge is critical. "Differently situated people may face different moral problems or experience similar ones differently" (Walker, 2007, p. 56). Using 'our' assumptions about what is moral can easily render others mute, incoherent, and unacknowledged. As a result, questions about the identification and interpretation of moral problems and the plausibility of moral claims must be addressed. We must critically interrogate the situated character of our own claims about what is moral.

Moral knowledge must be justified. Justification does not come from principles. "The only thing that corrects or refutes a morality on moral grounds is another, better-justified morality that shows the first one is wrong. What is involved in justifying a morality, however, is no one thing and no simple ones" (Walker, 2007, p. 238). We start from "from some moral perspective to know whether we are going to be talking about benefits and costs, or about tolerability and limiting conditions, or excellences or flaws, or reverence and defilements" (Walker, 2007, p. 249). Once we have oriented ourselves, we ask: "What is this way of life. . . ?, What good comes of this way of life . . . .?, [and] What can be said for it?" (Walker, 2007, p. 248). The questions are asked with an eye toward seeing whether the practices withstand scrutiny and whether they are worth continuing to endorse. It is good to look for instabilities because looking for situations where people do not yet understand how their lives impact the lives of others "is one of the most tractable and fruitful places for moral inquiry to begin" (Walker, 2007, p. 250). Justification ends, however tentatively and loosely, with the "habitability and acceptability of the common life to which they lead" (Walker, 2007, p. 77).

Ethical deliberation proceeds *via* narratives rather than principles. We need to know who is involved, how they are involved, how they understand their involvement, how they are related to each other, and how society and institutions frame their options. We progressively and mutually adjust and acknowledge the concerns. This is a kind of reflective equilibrium, but one that involves equilibrium among people who have limited knowledge of the workings of the moral-social order. Since I am interested here in sustainability, I will add limited knowledges of the natural order and of the interactions between the social and natural orders to that claim. Morality is an ongoing project.

## The Expressive-Collaborative Model in Sustainability

In an expressive-collaborative approach to ethical deliberation about sustainability, we constantly limn the geographies of our responsibilities, critically interrogating the assumptions that give rise to those responsibilities and attempting to justify new responsibilities and actions as better, more habitable than the status quo. I illustrate this by sketching examples from biodiversity conservation and livestock farming. The examples call attention to place, praxis, positionality, power, and context. The explicit attention to ethical justification adds to examinations of the link between sustainability and social justice, an underdeveloped area of inquiry (Sze, 2018).

A few words of qualification about the examples are in order. I highlight only matters of ethical deliberation, deliberately avoiding many of the larger systemic issues they raise. Also, I am not a member of the communities in the examples and so comment as an outsider. My nearest connection is to livestock farming. My family had a small herd of beef cattle until the 1980s. The pastureland that I now rent to a family friend is used as feed for his beef cattle, and some crops produced on the arable land that I now rent to him and another friend inevitably make their way into the livestock farming system. Since I do not experience directly the responsibilities and obligations involved in the examples, I can at best argue that the communities are asking "What is this way of life?, What good comes of it?, and what can be said for it?" in a naturalized, critical and justificatory fashion.

Practices of biodiversity preservation offer rich sources to illustrate the expressive-collaborative approach. Western conceptions of preservation are often in tension with livelihoods, as when forest preserves limit or exclude use of the forest by local communities. Increasingly, proponents of preservation recognize that such matters involve biophysical and cultural realms and their complex interaction (McCarter et al., 2018; Perfecto et al., 2019; Sarkar & Montoya, 2011; Walter & Hamilton, 2014).

As a specific example, consider McCarter et al.'s (2018) process of creating sustainability indicators that reflect place-based social and cultural contexts in the Western Province of the Solomon Islands. Melanesia in general is recognized as a highly threatened biodiversity hotspot. International guide-lines for resource management such as the Convention on Biological Diversity and the Sustainable Development Goals often prioritize Western worldviews of biodiversity, preservation, sustainability, and well-being. An example in the guidelines is the explicit standard of having a certain amount of land in a pro-tected area. These standards and goals are often at odds with local norms for resource use. McCarter's research team worked alongside four local commu-nities to identify indicators related to place-based conceptions of well-being. The goal was to help the communities plan for conservation work. The four sites have different languages, livelihoods (involving the sale of agricultural products, fish and/or handicrafts), religious affiliations, and governance struc-tures. In visioning sessions and interviews, 18 wide-ranging components of well-being were identified. They included: the maintenance of healthy marine resources, having productive and accessible gardens, the presence and use of locally important agro-biodiversity, healthy forests, transmission and use of vernacular languages, the presence of viable income generating activities, and legitimate and functional resource management bodies. Methodologically, the research team worked to minimize power differentials between place-based participants and the researchers. In addition to reason-giving, visual methods such as drawing, mapmaking, and photography were used to gather informa-tion about the dimensions of well-being. This allowed participation by sec-tions of the population who are often excluded in such discussions.

This process appears (it is still somewhat new) to allow a natural, critical, and justificatory approach to biodiversity conservation and the larger sustain-ability issues it relates to. Instead of relying on culturally and theoretically distant notions of well-being as it relates to biodiversity preservation, mem-bers of local communities in conjunction with the researchers characterized place-based dimensions of well-being. It is critical in at least two senses. The researchers helped the members voice the dimensions of well-being. They did not assume that members were already aware of all the dimensions of well-being and their interactions. Also, it was attentive to differently situated people and the different livelihoods of the different communities, using mul-tiple modes of input beyond reason-giving. And it focused on justifying the components of well-being as significant, as matters of ultimate concern for the communities involved. In doing so, it offers a better-justified morality that shows how the priorities of external NGOs and larger governance structures do not address the conception of well-being voiced by the communities. That justification is a multitude of things, each of which is itself complexly related to their biophysical and cultural realms.

As another example, consider livestock farming. Many view such prac-tices as inherently unsustainable. In their various forms, they produce green-house gas emissions, sometimes contribute to increased antibiotic resistance, sometimes violate workers' rights, raise concerns about animal welfare, etc. However, Rademaker et al. (2017) argue that if we focus on the identity of the practices involved, some forms of livestock farming can be ethically justified. Starting with the identity of practices means considering how farmers use resources of land and water and how they shape the animals and the environment. It also means considering that livestock farming is not just about keeping animals; it is about making products. This economic aspect of the identity is not just a matter of efficiency. "Valuable" means a much broader range of benefits, including but not limited to providing different products to society and keeping the farm going. This identity is associated with norms: jural norms involving the just treatment of employees and fair trade with customers and suppliers; aesthetic norms involving questions about how animals, meadows and water sources look; ethical norms related to the conditions for workers and conditions under which animals are kept; and many others. The question of identity is also bound up with the food systems of which any given livestock farm is a part. Each aspect of the identity is associated with standards of excellence that characterize good performance of the practices that constitute the identity.

Now suppose someone thinks livestock farming does not live up to the moral ideal of sustainability. They appeal to the environmental, human, or animal harms involved. In an expressive-collaborative approach, such an appeal does not by itself settle the question. Rather, we ask: what is this way of life?, what good comes of it?, and what can be said for it?. These questions are answered with reference to the identities associated with the practices. If practices of husbandry for chickens impede, say, their abilities to dust bathe, then one can develop a claim that this way of life is not habitable, for the chickens, us and/or the environment. To be sure, this does not settle the matter. But the "geographies of responsibilities" (Walker, 2007, p. 105) can begin to be charted. As much as it possible from our limited points of view, we query all systemic features of the practice. And we do not have to see ourselves as offering just a rebuke. We can justify the claim by pointing to ways the policy leads to the undoing of the larger practices: the chickens suffer physically; they are treated instrumentally, as mere stock; the more frequent restocking required by the policy might compel the farmer to have a larger supply of ready cash or to secure a loan; there may be environmental consequences associated with this behavior; etc. If we adopt a fully critical perspective, all aspects of the identity of the practice are on the table. In this instance, the judgment closely tracks those made from theoretical-juridical perspectives about animal welfare, but the range of social, ecological, and socio-ecological considerations invoked is much larger.

I am reporting the discussions third-hand, so it is hard to get a visceral sense of the responsibilities and obligations and their ties to identities. Interviews can provide some sense of this. As reported by Reo et al. (2019, p. 218), one interviewee recalled an Inuit Elder stating that being forced to stop using their language was like "having our tongues ripped out." Even without interview excerpts, we can say that the deliberations involve more than preference satisfaction, in monetary or other value terms. Livelihoods that involve cultural identities are important. Even in the case of livestock farming, 'valuable' relates to the farm's relation to society and to how practices keep the farm going. We can also say that it involves more specificity than consideration of ecological, throughput and socio-political sustainability considered through the lenses of respect for others and care. Matters of ideology, power and identity are involved. These are involved in the justifications of which dimensions of sustainability will be highlighted and which aspects of well-being will be traded off against each other. At least for the Solomon Islands example, the deliberations involve more than inviting everyone to the table and having them offer reasons once there. They involve awareness of power and positionality regarding who sits at the table and how those sitting express their concerns. The deliberations appeal to blame- and praiseworthiness, loss or preservation of identities, etc. They test whether what we feel and think is fitting. They test whether those fits deserve their authority over us. Self-reflexive learning about the consequences of actions and beliefs involves adjustments to personal and community identities. With respect to Thompson's virtue theory, it involves a critical stance toward functional integrity and the ideals of citizenship, dependence on local others and self-realization. Specifically, questions about which integrity is desired and who contributes to that integrity are central to the discussions. Who gets left out in the original formulations of a good, virtuous sustainability? Why is it important to pay attention to these claims? These must be justified in the context. We learn to express the commitments and responsibilities we have and what can be changed about them. We enter into the discussion with a partial awareness of what we and others want, and we push and pull, always being cognizant that our expressions may not be fully representative of even our own values and interests much less those of others in the community and those who are not in the community.

## Notes

1  Walker (2007, p. 43) notes another sense of juridical, in which moral theories are judged on their logical and epistemological adequacy. This sense is not relevant to my discussion.
2  Thanks to my colleague Susan Levin for enlightening me about matters of Ancient Greek philosophy.
3  Thanks to my colleague Alex Barron for enlightening me about matters of governmental policy-making.
4  Norton could respond that his conception of reason giving in the public realm encompasses storytelling and other forms of discourse. This seems viable given that, for Norton, all discourse is a mix of description and prescription. However, Norton (2015, pp. 218–257) focuses on how processes of deliberation and public discussion allow a new multigenerational public interest to emerge. Individuals express concerns, and this allows a community consensus in goals and values. Little is said about the form of testimony.

## References

Addelson, K. (1994). *Moral passages: Toward a collectivist moral theory.* Routledge.

Allen, A. (2017). Power/knowledge/resistance: Foucault and epistemic injustice. In I. Kidd, J. Medina & G. Pohlhaus, Jr. (Eds.), *The Routledge handbook of epistemic injustice* (pp. 187–194). Routledge.

Allhoff, F. (2010). What are applied ethics? *Science and Engineering Ethics, 17*(1), 1–19. https://doi.org/10.1007/s11948-010-9200-z

Annas, J. (1981). *An introduction to Plato's republic.* Clarendon.

Arney, R. N., Henderson, M. B., DeLoach, H. R., Lichtenstein, G., & German, L. A. (2022). Connecting across difference in environmental governance: Beyond rights, recognition, and participation. *Environment and Planning E: Nature and Space, 6*(2), 251484862211088. https://doi.org/10.1177/25148486221108892

Battin, M. P., Francis, L. P., Jacobson, J. A., & Smith, C. B. (2009). *The patient as victim and vector: Ethics and infectious disease.* Oxford University Press.

Bevacqua, M., & Bowman, I. (2018). I Tano' I Chamorro/Chammora Land: Situating sustainabilities through spatial justice and cultural perpetuation. In J. Sze (Ed.), *Sustainability: Approaches to evnironmental justice and social power* (pp. 196–221). New York University Press.

Chakraborty, D., Hardman, S., Karten, S., & Tal, G. (2021). *No, electric vehicles aren't driven less than gas cars.* University of California, Davis Institute of Transportation Studies. https://its.ucdavis.edu/blog-post/no-electric-vehicles-arent-driven-less-than-gas-cars/

Cruz-Torres, M. L., & McElwee, P. (2012). *Gender and sustainability: Lessons from Asia and Latin America.* University of Arizona Press.

Curren, R., & Metzger, E. (2017a). Preserving opportunity: A précis of living well now and in the future: Why sustainability matters. *Ethics, Policy & Environment, 20*(3), 227–239. https://doi.org/10.1080/21550085.2017.1374000

Curren, R., & Metzger, E. (2017b). *Living well now and in the future: Why sustainability matters.* MIT Press.

De Brasi, L., & Warman, J. (2023). Deliberative democracy, epistemic injustice, and epistemic disenfranchisement. *Logos & Episteme, 14*(1), 7–27. https://doi.org/10.5840/logos-episteme20231411

Domínguez, L., & Luoma, C. (2020). Decolonising conservation policy: How colonial land and conservation ideologies persist and perpetuate indigenous injustices at the expense of the environment. *Land, 9*(3), 65. https://doi.org/10.3390/land9030065

Fraser, N. (1990). Rethinking the public sphere: A contribution to the critique of actually existing democracy. *Social Text, 25/26*, 56–80. https://doi.org/10.2307/466240

Fricker, M. (2007). *Epistemic injustice: Power and the ethics of knowing.* Oxford University Press.

Geels, F. W. (2012). A socio-technical analysis of low-carbon transitions: Introducing the multi-level perspective into transport studies. *Journal of Transport Geography, 24*, 471–482. https://doi.org/10.1016/j.jtrangeo.2012.01.021

Hayden, R., Kaye, A., Masur, K., Miller, S., O'Donovan, S., Rowland, L., & West, S. (2013). *FREEDOM: A documentary history of emancipation, 1861–1867, Series 3, Volume 2- Land And Labor; 1866–1867.* University of North Carolina Press.

Jupiter, S. (2017). Culture, kastom and conservation in Melanesia: What happens when worldviews collide? *Pacific Conservation Biology, 23*(2), 139–145. https://doi.org/10.1071/pc16031

Kaaria, S., & Osorio, M. (2018). *The gender gap in land rights*. FAO. https://www.fao.org/3/I8796EN/i8796en.pdf

Kashwan, P., Duffy, R. V., Massé, F., Asiyanbi, A. P., & Marijnen, E. (2021). From racialized neocolonial global conservation to an inclusive and regenerative conservation. *Environment: Science and Policy for Sustainable Development*, *63*(4), 4–19. https://doi.org/10.1080/00139157.2021.1924574

Kohn, M. (2000). Language, power, and persuasion: Toward a critique of deliberative democracy. *Constellations*, *7*(3), 408–429. https://doi.org/10.1111/1467-8675.00197

Kothari, U. (2001). Power, knowledge and social control in participatory development. In B. Cooke & U. Kothari (Eds.), *Participation: The new tyranny?* (pp. 139–152). Zed Books.

Kurz, T., Gardner, B., Verplanken, B., & Abraham, C. (2014). Habitual behaviors or patterns of practice? Explaining and changing repetitive climate-relevant actions. *WIREs Climate Change*, *6*(1), 113–128. https://doi.org/10.1002/wcc.327

Mayumi, K., & Giampietro, M. (2006). The epistemological challenge of self-modifying systems: Governance and sustainability in the post-normal science era. *Ecological Economics*, *57*(3), 382–399. https://doi.org/10.1016/j.ecolecon.2005.04.023

MacIntyre, A. (1981). *After virtue*. Notre Dame Press.

McCarter, J., Sterling, E. J., Jupiter, S. D., Cullman, G. D., Albert, S., Basi, M., Betley, E., Boseto, D., Bulehite, E. S., Harron, R., Holland, P. S., Horning, N., Hughes, A., Jino, N., Malone, C., Mauli, S., Pae, B., Papae, R., Rence, F., … & Filardi, C. E. (2018). Biocultural approaches to developing well-being indicators in Solomon Islands. *Ecology and Society*, *23*(1), 32. https://doi.org/10.5751/es-09867-230132

McKenna, E. (2011). Feminism and farming: A response to Paul Thompson's the Agrarian Vision. *Journal of Agricultural and Environmental Ethics*, *25*(4), 529–534. https://doi.org/10.1007/s10806-011-9328-0

McShane, T. O., Hirsch, P. D., Trung, T. C., Songorwa, A. N., Kinzig, A., Monteferri, B., Mutekanga, D., Thang, H. V., Dammert, J. L., Pulgar-Vidal, M., Welch-Devine, M., Peter Brosius, J., Coppolillo, P., & O'Connor, S. (2011). Hard choices: Making trade-offs between biodiversity conservation and human well-being. *Biological Conservation*, *144*(3), 966–972. https://doi.org/10.1016/j.biocon.2010.04.038

McShane, T. O., & Newby, S. A. (2004). 4. Expecting the unattainable: The assumptions behind ICDPs. In T. McShane & M. Wells (Eds.), *Getting biodiversity projects to work* (pp. 49–74). Columbia University Press.

Merino-Saum, R. (2018). Re-politicizing participation or reframing environmental governance? Beyond indigenous' prior consultation and citizen participation. *World Development*, *111*, 75–83. https://doi.org/10.1016/j.worlddev.2018.06.025

Miller, T. R., Wiek, A., Sarewitz, D., Robinson, J., Olsson, L., Kriebel, D., & Loorbach, D. (2013). The future of sustainability science: A solutions-oriented research agenda. *Sustainability Science*, *9*(2), 239–246. https://doi.org/10.1007/s11625-013-0224-6

Mitchell, R. (2022). The energy historian who says rapid decarbonization is a fantasy. *Los Angeles Times*. https://www.latimes.com/business/story/2022-09-05/the-energy-historian-who-says-rapid-decarbonization-is-a-fantasy

National Research Council. (2013). *Sustainability for the nation: Resource connections and governance linkages*. The National Academies Press.

Norton, B. (2005). *Sustainability: A philosophy of adaptive ecosystem management*. University of Chicago Press.

Norton, B. (2015). *Sustainable values, sustainable change: A guide to environmental decision making*. University of Chicago Press.

O'Neill, J., Holland, A., & Light, A. (2008). *Environmental values.* Routledge.

Parmar, H., Chandramowli, S., & Boddu, D. (2023). *Selecting the right clean energy project sites can maximize IRA benefits.* ICF. https://www.icf.com/insights/energy/clean-energy-project-sites-IRA-benefits.

Perfecto, I., Vandermeer, J., & Wright, A. (2019). *Nature's matrix: Linking agriculture, biodiversity conservation and food sovereignty.* Routledge.

Rademaker, C. J., Glas, G., & Jochemsen, H. (2017). Sustainable livestock farming as normative practice. *Philosophia Reformata, 82*(2), 216–240. https:// doi 10.1163/23528230-08202002

Reo, N. J., Topkok, S. M., Kanayurak, N., Stanford, J. N., Peterson, D. A., & Whaley, L. J. (2019). Environmental change and sustainability of indigenous languages in Northern Alaska. *ARCTIC, 72*(3), 215–228. https://doi.org/10.14430/arctic68655

Robinson, J. (2004). Squaring the circle? Some thoughts on the idea of sustainable development. *Ecological Economics, 48*(4), 369–384. https://doi.org/10.1016/j.ecolecon.2003.10.017

Sarkar, S., & Montoya, M. (2011). Beyond parks and reserves: The ethics and politics of conservation with a case study from Perú. *Biological Conservation, 144*(3), 979–988. https://doi.org/10.1016/j.biocon.2010.03.008

Schneider, S. H. (1997). Integrated assessment modeling of global climate change: Transparent rational tool for policy making or opaque screen hiding value-laden assumptions? *Environmental Modeling & Assessment, 2,* 229–249.

Seto, K. C., Davis, S. J., Mitchell, R. B., Stokes, E. C., Unruh, G., & Ürge-Vorsatz, D. (2016). Carbon lock-in: Types, causes, and policy implications. *Annual Review of Environment and Resources, 41*(1), 425–452. https://doi.org/10.1146/annurev-environ-110615-085934

Shove, E. (2010). Beyond the ABC: Climate change policy and theories of social change. *Environment and Planning A: Economy and Space, 42*(6), 1273–1285. https://doi.org/10.1068/a42282

Simpson, L. (2017). *As we have always done: Indigenous freedom through radical resistance.* University of Minnesota Press.

Sze, J. (Ed.) (2018). *Sustainability: Approaches to environmental justice and social power.* New York University Press.

Sze, J. (2020). *Environmental justice in a moment of danger.* University of California Press.

Taylor, C. (1989). *Sources of the Self.* Harvard University Press.

Thompson, P. B. (2010). *The Agrarian vision.* University Press of Kentucky.

Thompson, P. B. (2018). Norton and sustainability as such. In S. Sarkar & B. Minteer (Eds.), *A sustainable philosophy—The work of Bryan Norton* (pp. 7–26). Springer.

Valera, L., Vidal, G., & Leal, Y. (2020). Beyond application. The case of environmental ethics. *Tópicos, Revista de Filosofía, 60,* 437–460. https://doi.org/10.21555/top.v0i60.1122

Walter, R. K., & Hamilton, R. J. (2014). A cultural landscape approach to community-based conservation in Solomon Islands. *Ecology and Society, 19*(4), 41. https://doi.org/10.5751/es-06646-190441

Walker, M. U. (2007). *Moral understandings: A feminist study in ethics,* 2nd ed. Oxford University Press.

WCED (World Commission on Environment and Development) (1987). *Our common future.* Oxford University Press.

Wellstead, A., Howlett, M., & Rayner, J. (2016). Structural-functionalism redux: Adaptation to climate change and the challenge of a science-driven policy agenda. *Critical Policy Studies, 11*(4), 391–410. https://doi.org/10.1080/19460171.2016.1166972

Whyte, K., Caldwell, C., & Schaefer, M. (2018). Indigenous lessons about sustainability are not just for "all humanity. In J. Sze (Ed.), *Sustainability: Approaches to environmental justice and social power* (pp. 149–179. New York University Press.

Williams, G. (2004). Evaluating participatory development: Tyranny, power and (re) politicisation. *Third World Quarterly, 25*(3), 557–578. https://doi.org/10.1080/0143659042000191438

Williams, B. (1981). *Moral luck.* Cambridge University Press.

Woodside, C. (2016). *Libertarians on the prairie.* Arcade.

Young, I. M. (1990). *Justice and the politics of difference.* Princeton University Press.

Young, I. M. (2000). *Inclusion and democracy.* Oxford University Press.

# 6    Conclusion

I have argued that many mainstream and some alternative approaches to sustainability assume a folk conception of science and its relation to society. One commonly finds explicit and implicit use of referentialist theories of meaning, views of theories as compact sets of scientific and moral propositions that sanction strong inferences, and simple or positivist notions of evidence. If change is seen as needed to aid in the pursuit of sustainability, these are usually left unchallenged and unchanged. Instead, changes are made to the substantive elements of sustainability: the three pillars are replaced with planetary boundaries or a thermodynamic framework; strong sustainability becomes normative sustainability; resource sufficiency becomes functional integrity; new indicators are developed with new empirical evidence; etc. But these moves change only the *substance* of the proposals, leaving the *nature* of the concepts, theories, evidence, and ethical commitments largely unchallenged and unchanged.

I have argued that the nature of the elements in the folk conception constrain our attempts to pursue sustainability. Other social, economic, and political factors are also surely responsible for our lack of progress, but the folk conception is at least partly responsible, probably because it feeds into dominant, business-as-usual ways of thinking. To challenge the folk conception, we need to change both our conception of science and its relation to society and the tools we use to critically analyze those conceptions.

I have tried to argue for the inadequacy rather than the falsity of the 'folk' conception of science and its relation to society. In the complex, wicked messes of sustainability, it is inapposite and unhelpful. In these messes, we need a view of science in which: knowledge is partial and situated; there are different legitimate factual and normative representations of the problems; trade-offs make objectively optimal solutions difficult or impossible; and the problem situations themselves shift as we intervene in them. Recognizing these phenomena lets us rejigger the folk conception. This reworking has involved appealing to recent perspectives in the philosophy of science that reveal the restrictions of the folk conception and provide more expansive platforms for analysis.

DOI: 10.4324/9781003268697-6

We enrich the pursuit of sustainability by developing notions of meaning, inferential reasoning, evidence, and ethical decision-making that appeal to the social embeddedness of practices. To understand what sustainability refers to and whether an action deserves to be called sustainable, we must start from the ground up and give reasons why a particular proposal 'plays the game' of sustainability. Deciding that it does or does not, that it is or is not an acceptable extension, is always done against a backdrop of a previous network of interpretations. To get a theoretical grasp on sustainability, we appeal to the patchwork of projected and extended meanings for sustainability. Inferences are justified materially, i.e., as relations of support based on the specific subject matter. Indicators are complex empiricist models connected essentially to their contexts because they are assumption- and theory-laden. To articulate the moral ideal of sustainability, we start with a view that sees moral knowledge as produced and sustained within communities. We attend to the geography of responsibilities and obligations in the community, always testing whether current moral understandings make sense for others and whether they deserve their authority over us. All these moves make us reflect more fully on the local, place-based nature of sustainability claims. In so doing, they at least point in the direction of addressing matters of social, political, institutional, and cultural power; every claim must be scrutinized in relation to context. If we can highlight ways in which contextual evidence, theory and ethical framing are being ignored, we are better prepared to alter the dominant economic reading of sustainability. If we can think in these terms, we can provide a firmer, more justified ground for claims that a given proposal advances our pursuit of sustainability, broadly construed.

I have provided no arguments (and have no illusions) that appealing to socially embedded practices will change the mindsets and behaviors of those who, for whatever reason, resist the pursuit of sustainability. But as I indicate below, I do think it is possible to challenge them. 'Business as usual' becomes harder if one questions the way assumptions condition theories, evidence, inferences, and moral claims.

As I have tried to indicate throughout the book, the argument resonates with existing views in sustainability discourse. Sustainability scientists, particularly those who see sustainability as an essentially transdisciplinary endeavor involving matters beyond science, as well as action-oriented researchers and activists, have pointed in a similar direction. The philosophical rationale undergirds their scientific and socio-political strategies. And it provides ways of detecting when proposed changes continue to use notions that constrain the effort.

Why might it matter to think about sustainability in this way? For philosophers of science and for sustainability scholars and practitioners intrigued by Nagatsu et al.'s (2020) claim that philosophers of science can contribute to the development and soundness of sustainability science, it opens questions about the nature of inquiry in the philosophy of science. In terms of Plaisance and Elliott's (2021) framework that outlines different types of socially engaged philosophy of science, I have worked individually toward a view of sustainability that is very epistemically integrated. The individual aspect is, I think, obvious. In terms of epistemic integration, I have incorporated conceptions of meaning, theoretical inference, evidence, and normativity drawn from the philosophy of science into sustainability discourse. This led to the claim that mainstream sustainability discourse should change its epistemic and normative practices. I have also incorporated sustainability discourse into the philosophy of science. Sustainability forces us to adopt more complex, place-specific conceptions of meaning, inference, evidence, and ethical deliberation. And it forces us to alter the aim of philosophical inquiry. Rather than revealing misunderstandings about theoretical inferences (including those associated with hypothesis testing and confirmation) and the nature of evidence in order to rebut those who try to undermine scientific consensus (Cartieri & Potochnik, 2013), the point is to construct accounts of epistemic and methodological practices that are sensitive to the nature of the situation. For sustainability, we cannot lift the veil of misunderstanding, revealing problematic appeals to accepted, consensus-based constructions of the problem. There is little consensus (other than a high-level one that something is wrong). In wicked situations, there are multiple legitimate constructions of a problem. A few readings might be able to be eliminated through the invocation of testing, evidential standards and such, but many will remain. We should be careful not to analyze sustainability (and other wicked problems) with the philosophical assumptions embedded in the folk view of science.

I do not have a complete sense of this kind of engagement. However, it does involve using philosophical inquiry to change the understanding of the situation, and it does involve using the situation to inform the choice of philosophical tools and the aim of philosophical inquiry (B. Burkhart, personal communication, December 2, 2022). Rather than reading a problem through pre-existing categories, it asks what experiences matter in the situation. If the situation requires a recasting of what the philosophical issues are and how they are understood, we undertake that project. Notions of meaning, inferential structure, evidence, and the relation of science to social and ethical issues are altered. This looks something like the way philosophy is used in field philosophy, but I think success here is not measured only by crafting a new "policy, practice, community or object" (Brister & Frodeman, 2020, p. 5). Success is a new understanding of the nature of sustainability and its links to policies and society.

Lloyd's (2012) call for philosophers of science to pay attention to changing methodologies in the sciences, particularly climate science, is relevant here. When adopting a complex empiricist understanding of models and data, philosophers of science must change their conceptions of scientific processes and reasoning. Inferences from theory-laden hypotheses, models, and evidence inherently involve reference to background assumptions and their validity. Rather than assuming traditional tools are adequate, they are changed to fit the situation at hand.

Another example comes from medical ethics, a field that, like sustainability, juggles complex mixes of empirical and ethical considerations. Francis (2016) relates how epidemics were read through the lens of autonomy during the 1980s and 1990s. HIV/AIDS and the other epidemics that followed were initially approached through ethical frameworks centering on issues of informed consent and confidentiality. Increasingly, it became clear these tools were insufficient and incomplete. New (or recovered) ethical issues involved in planning for outbreaks, justice in primary care, justice for first responders and essential personnel, and public health measures now became important. For the AIDS epidemic, racial, and social issues, which had been largely missing from the discussion, emerged as central questions. Again, the situation altered the conception of the tools needed for philosophical inquiry.

For sustainability, applying traditional tools and frameworks from the philosophy of science, ones tied to the folk view of science, frames the analysis in restrictive ways and misses several important issues. The relation of theories, inferences, evidence, and moral responsibility to their context is overlooked. Matters of power, place and position get elided. To get at these, we need to reconfigure the process of meaning-making, inference-making, evidence-making, and ethical deliberation. These need to be part of the philosophical discourse of sustainability and of sustainability discourse more generally.

What does the book's argument imply for scholars and practitioners of sustainability? It relativizes claims of sustainability to their local context. As with nearly all the rest of my claims about sustainability throughout the book, this is not a novel claim. But many mainstream approaches do search for universal frameworks. The book's argument reinforces the message of local context, providing philosophical rationales for resisting the substantive, methodological and philosophical impulses that lead toward universal conceptualizations of sustainability. As much as universal frameworks look appealing, science as well as politics is local. In Chapter 4, I invoked Winner's (1980) argument that artifacts have politics. By analogy, sustainability claims – including those based in science – have politics. We need to look at what sustainability claims do, not just what they mean (on a traditional conception of meaning that is independent of practice).

Clearly, I have not substantiated a claim that seeing sustainability as involving model-based, normative claims about wicked problems can be used to intervene in current discussions about sustainability. Those hoping for clear policy proposals will by now surely be disappointed. However, this is a work of philosophy. I hope that clarifying concepts, formulating new perspectives on concepts, theories, evidence, and ethical deliberation, and critiquing the ontological, epistemological, and ethical assumptions of any given proposal (LaPlane et al., 2019) can set the stage for more "complexity-oriented, critically reflexive, and normatively committed" (West et al., 2019, p. 536) inquiries into sustainability.

We can criticize business-as-usual environmental policy proposals that prioritize economic growth and efficient resource use and downplay or ignore local conceptions of well-being. These continue to win out in plural, (supposedly) democratic planning discussions and processes (Norton, 2015; Pollans, 2019; Thompson, 2018). They are often skewed in favor of those who have resources and standing. Decisions that favor them are made through a blend of instinct and influence rather than through 'rational,' 'reasoned' processes (Pollans, 2019; Shrader-Frechette, 2013; Stone, 2012). Beginning with claims that are partial, situated in place and time, shot through with power and privilege, we can criticize more fully proposals claimed to be neutral. We can stimulate consideration of alternative points of view by recognizing our own and others' positionality and limitations.

As well, we can question the framings of sustainability. When we are critically reflexive, we interrogate our scientific claims, our commitments to political institutions and processes, our responsibilities, and our emotions. For example, as we in the industrialized North respond to the urgent problems of sustainability, the dominant affective modes are despair and nihilism, at least in the case of climate change (Atkin, 2017; Fiskio, 2017; Piper, 2022). Fridays for the Future, a youth-led and -organized global school strike movement inspired by Greta Thunberg's actions, has asked "Why study for a future, which may not be there?" (Wackernagel, 2019). A critically reflexive approach can help "curb your catastrophism" (Skrimshire, 2008). As climate scientist Michael Mann says,

> I am not a fan of this sort of doomist framing. It is important to be up front about the risks of unmitigated climate change, and I frequently criticize those who understate the risks. But there is also a danger in overstating the science in a way that presents the problem as unsolvable, and feeds a sense of doom, inevitability and hopelessness.
>
> (quoted in Atkin, 2017)

As we act to address the problems, it can be useful to think and reflect on the ways we construct the issues. Seeing problems as catastrophic often overlooks their chronic nature, and it often calls forth state-level technological and scientific solutions (Wester et al., 2022). As the saying goes: marry in haste, repent at leisure. Mixing metaphorical proverbs, far better to 'measure twice (seven times if you are Russian) and cut once.' Rather than taking despair, nihilism, and urgency for granted, we can reflect how affectual and emotional responses – as well as understandings of factual matters – are shaped by "lived realities of class, nation, and community" (Sze, 2020, p. 82). Who are we such that we have this response? What geography of responsibilities allows us to construct and reproduce these feelings and to deliver a judgment that they are appropriate? What do these feelings do to, with and for differently placed people and the natural processes that they relate to? Asking these questions can allow us to follow many environmental justice scholars and practitioners who "reject despair" and opt for "principled resistance, life affirmation . . ., and solidarities based on radical empathy, humor, grace, and transformation" (Sze, 2020, p. 80). Urgency, yes; apocalypse now, no.

To end by again invoking Wittgenstein, thinking about sustainability from an engaged standpoint means addressing fully the normative structure of the scientific, social, and political backgrounds of sustainability claims. We show how the techniques of the dominant discourse block change, and we show how the techniques of newer discourses allow justification of better conceptions of social and natural well-being. And then we display mastery of the techniques, teaching others to go on in this way. This justifies them. This is a slower process than many might want. But the aim is to reveal the commitments that allow certain kinds of normative judgments and reactions to be seen as 'natural,' to question them, and to replace them with other commitments that we deem better for our well-being.

## References

Atkin, E. (2017). The power and peril of 'climate disaster porn.' *New Republic.* https://newrepublic.com/article/143788/power-peril-climate-disaster-porn

Brister, E., & Frodeman, R. (Eds.) (2020). *A guide to field philosophy: Case studies and practical strategies.* Routledge.

Cartieri, F., & Potochnik, A. (2013). Toward philosophy of science's social engagement. *Erkenntnis, 79*(S5), 901–916. https://doi.org/10.1007/s10670-013-9535-3

Fiskio, J. (2017). Building paradise in the classroom. In S. Siperstein, S. Hall & S. LeMenager (Eds.), *Teaching climate change in the humanities* (pp. 101–109). Routledge.

Francis, L. (2016). Applied ethics: A misnomer for a field? *Proceedings and Addresses of the American Philosophical Association, 90*, 40–54. http://www.jstor.org/stable/26622937

Laplane, L., Mantovani, P., Adolphs, R., Chang, H., Mantovani, A., McFall-Ngai, M., ... & Pradeu, T. (2019). Opinion: Why science needs philosophy. *Proceedings of the National Academy of Sciences, 116*(10), 3948–3952. https://doi/10.1073/pnas.1900357116

Lloyd, E. A. (2012). The role of 'complex' empiricism in the debates about satellite data and climate models. *Studies in History and Philosophy of Science Part A, 43*(2), 390–401. https://doi.org/10.1016/j.shpsa.2012.02.001

Nagatsu, M., Davis, T., DesRoches, C. T., Koskinen, I., MacLeod, M., Stojanovic, M., & Thorén, H. (2020). Philosophy of science for sustainability science. *Sustainability Science, 15*, 1807–1817. https://doi.org/10.1007/s11625-020-00832-8

Norton, B. (2015). *Sustainable values, sustainable change: A guide to environmental decision making.* University of Chicago Press.

Piper, K. (2022). Stop telling kids that climate change will destroy their world. *Vox.* https://www.vox.com/23158406/climate-change-tell-kids-wont-destroy-world

Plaisance, K. S., & Elliott, K. C. (2021). A framework for analyzing broadly engaged philosophy of science. *Philosophy of Science, 88*(4), 594–615. https://doi.org/10.1086/713891

Pollans, L. B. (2019). Sustainability policy paradox: Coping with changing environmental priorities in municipal waste management. *Journal of Environmental Policy & Planning, 21*(6), 785–796. https://doi.org/10.1080/1523908x.2019.1673157

Shrader-Frechette, K. (2013). [Review of the book *Unsimple Truths*, by S. Mitchell]. *The British Journal for the Philosophy of Science, 64*(2), 449–453. https://doi.org/10.1093/bjps/axs023

Skrimshire, S. (2008). Curb your catastrophism. *Red Pepper.* www.redpepper.org.uk/curb-your-catastrophism.

Stone, D. (2012). *Policy paradox: The art of political decision-making*, 3rd ed. W. W. Norton.

Sze, J. (2020). *Environmental justice in a moment of danger.* University of California Press.

Thompson, P. B. (2018). Norton and sustainability as such. In S. Sarkar & B. Minteer (Eds.), *A sustainable philosophy—The work of Bryan Norton* (pp. 7–26). Springer.

Wackernagel, M. (2019). 'Why study for a future, which may not be there?' *Global Footprint Network.* https://www.footprintnetwork.org/2019/03/10/why-study-for-a-future-which-may-not-be-there/

West, S., van Kerkhoff, L., & Wagenaar, H. (2019). Beyond "linking knowledge and action": Towards a practice-based approach to transdisciplinary sustainability interventions. *Policy Studies, 40*(5), 534–555. https://doi.org/10.1080/01442872.2019.1618810

Wester, J., Turffs, D., McEntee, K., Pankow, C., Perni, N., Jerome, J., & Macdonald, C. (2022). Agriculture and downstream ecosystems in Florida: An analysis of media discourse. *Environmental Science and Pollution Research, 30*(2), 3804–3816. https://doi.org/10.1007/s11356-022-22475-1

Winner, L. (1980). Do artifacts have politics? *Daedalus, 109,* 121–136.

# Index

Note: Page numbers followed by "n" denote endnotes.

114 *Index*

For Product Safety Concerns and Information please contact our EU
representative GPSR@taylorandfrancis.com
Taylor & Francis Verlag GmbH, Kaufingerstraße 24, 80331 München, Germany

www.ingramcontent.com/pod-product-compliance
Lightning Source LLC
Chambersburg PA
CBHW061336220326
41599CB00026B/5215